In the eighteenth century the phenomenon of animal regeneration captured the attention and imagination of the era's leading naturalists and intellectuals. Their research on the phenomenon spurred the transition from descriptive natural history to modern experimental zoology. *A History of Regeneration Research* chronicles this crucial evolutionary stretch in the history of developmental biology and offers an insightful analysis of the milestones of regeneration research. The book not only discusses the leading researchers and their seminal discoveries in the field but also brings together critical commentaries on the social context and philosophical commitments that shaded their interpretations.

Opening up new ground, *A History of Regeneration Research* will be of great interest to scientists, as well as to historians of science.

A history of regeneration research

Oscar E. Schotté

This volume is dedicated to the memory of Oscar Emile Schotté (1895–1988), whose contributions to regeneration research both at the bench and through his students are a significant part of the history of this research field in America during the twentieth century. With intellectual roots that tapped the laboratories of Emile Guyénot, Hans Spemann, and Ross G. Harrison, Schotté will undoubtedly be prominent in future studies of the continuing history of regeneration research.

A history of regeneration research

Milestones in the evolution of a science

Edited by

CHARLES E. DINSMORE

Rush Medical College, Chicago

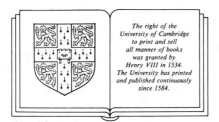

The right of the
University of Cambridge
to print and sell
all manner of books
was granted by
Henry VIII in 1534.
The University has printed
and published continuously
since 1584.

CAMBRIDGE UNIVERSITY PRESS

Cambridge
New York   Port Chester   Melbourne   Sydney

Published by the Press Syndicate of the University of Cambridge
The Pitt Building, Trumpington Street, Cambridge CB2 1RP
40 West 20th Street, New York, NY 10011, USA
10 Stamford Road, Oakleigh, Melbourne 3166, Australia

First published 1991

Printed in Canada

Library of Congress Cataloging-in-Publication Data
A history of regeneration research: milestones in the evolution of a
science / edited by Charles E. Dinsmore.
   p. cm.
Includes index.
ISBN 0-521-39271-3
1. Regeneration (Biology) – History.   I. Dinsmore, Charles E.
QH499.H57 1991
574.3'1'09 – dc20   90-26181
                                               CIP

British Library Cataloguing in Publication Data
A history of regeneration research.
1. Animals. Regeneration. Research, history
I. Dinsmore, Charles E.
591.31

ISBN 0-521-39271-3 hardback

Contents

Contributors

Keith R. Benson
Department of Medical History and
  Ethics
University of Washington, Seattle

Frederick B. Churchill
Department of the History and
  Philosophy of Science
Indiana University, Bloomington

John S. Cook
Biology Division
Oak Ridge National Laboratory,
  Tennessee

Charles E. Dinsmore
Department of Anatomy
Rush Medical College, Chicago

Jacqueline Géraudie
Laboratoire d'Anatomie Comparée
Université de Paris VII

Richard J. Goss
Division of Biology and Medicine
Brown University, Providence

Howard M. Lenhoff
Department of Developmental and
  Cell Biology
University of California, Irvine

Sylvia G. Lenhoff
Department of Developmental and
  Cell Biology
University of California, Irvine

Richard A. Liversage
Ramsay Wright Zoological
  Laboratories
University of Toronto, Canada

Jane Maienschein
Department of Philosophy
Arizona State University, Tempe

Marcus Singer
Department of Anatomy
School of Medicine
Case Western Reserve University,
  Cleveland

Dorothy M. Skinner
Biology Division
Oak Ridge National Laboratory,
  Tennessee

Joseph W. Vanable, Jr.
Department of Biological Sciences
Purdue University, West Lafayette,
  Indiana

Lewis Wolpert
Department of Anatomy and
  Developmental Biology
University College, London, and
  Middlesex School of Medicine

Preface

The life of a scientist is an odyssey that may begin with an individual's exposure to historically significant scientific discoveries. These, in turn, may stimulate an eager imagination, challenge the creative genius of the child within. They are the guideposts planted along the way by earlier adventurers that both indicate milestones achieved in clearing the wilderness of the unknown and illuminate pathways − that is to say, the process by which one arrives at scientific discovery. It seems to me that scientists, in the company of their colleagues in the history and philosophy of science, ought to reexamine with reasonable frequency the map of scientific progress in their chosen fields. In addition to refreshing or updating their own perspectives, another aim is to provide the apprentice with a useful tool for viewing the historical depth, breadth, and richness of the issues upon which the field was established. The product of that reflection and reevaluation, such as the proceedings of a symposium, should speed the interested explorer to the frontiers of the discipline. But it should do so by engendering an enthusiasm for developing a more profound understanding and appreciation of the foundations on which the discipline is built.

This book has several objectives. First, my desire to have access to a scholarly history of regeneration research is shared by many investigators, as I have discovered through conversations with colleagues over the years. Second, too often, although students of science are not introduced early to scholarly histories of science, they receive an abundance of anecdotal information − often erroneous (e.g., T. H. Morgan's reference, in his 1901 book *Regeneration*, to Abraham Trembley as an abbé) − that continues to be transmitted from teachers to students over the years. Therefore, the principal objective of this book is to create and provide a reference resource for scholars active in the field, as well as for present and future students of regeneration and its history.

ix

As a result of increasing specialization and the successes of scientific reductionism, researchers have been deciphering the language of the cells and dialogues of the tissues. Now, more than ever, it is important to step back and reflect upon the edifice of regeneration research to recapture the global, organismic significance of the molecular detail being added today at an accelerating pace. This is not just an academic exercise but, rather, a means by which we can better determine whether we have been faithful to the challenge of our predecessors. Have we answered legitimate questions about the fundamental nature of regeneration and its significance among the properties of living things? And in so doing, have we provided new insights and directions for future research? In what light will historians reflect upon our efforts? I hope this volume will stimulate the reader to think more deeply about the significance of regeneration research in the modern history of biology.

This book is derived largely from essays presented in a symposium on The History of Regeneration Research, held at the annual meeting of the American Society of Zoologists on December 30, 1988, in San Francisco. Its planning was inspired many years ago, as I began my own studies of amphibian limb regeneration, by my futile efforts to find a scholarly overview of the history of regeneration research. I am greatly in the debt of those friends and colleagues who, along the way, encouraged and supported my efforts to organize this project. My appreciation goes particularly to those who endured my editorial demands and whose research is represented in the following pages. If we have been successful in achieving our aims, it is to their great credit.

I was indeed fortunate to have benefited from the advice and guidance provided by Helen Wheeler and Sara-Jeanne Monsarrat of Cambridge University Press during the book's gestation. In addition, I thank Mary Adams-Wiley of the American Society of Zoologists, whose organizational skills and support were invaluable to the success of the symposium, without which this book might never have been born. Support for the symposium was provided, in part, by grants from Abbott Laboratories, Hoechst-Roussel, Monsanto Company, Roche Laboratories, and Rush–Presbyterian–St. Luke's Medical Center of Chicago.

1

Introduction

CHARLES E. DINSMORE

> If there were no regeneration, there could be no life. If everything
> regenerated there would be no death.
>
> R. J. Goss, *Principles of Regeneration* (1969), p. 1

Regeneration means many things to many people, as Richard Goss
illustrates in the following chapter by taking us on a tour, often colorful and always enlightening, of the idea's historical roots in myth,
anecdote, and science. One is asked to reflect not only on the scientific
reality of discoveries about the fundamental nature of regenerative
phenomena, their biology buried in the phylogenetic history of genes,
but also on the realm of naive human aspiration from an immemorial
imagination, a creative wellspring of curiosity and speculation. From
these has arisen an imposing list of as yet unanswered scientific questions concerning animal regeneration. That makes it necessary to
define, classify, and arbitrarily select those regenerative phenomena,
associated events, and individuals whose histories will be discussed
in this overview. My own interest in regeneration research has been in
trying to understand the mechanisms by which the restoration of complex form is regulated, as one sees in salamander limb regeneration.
That interest has clearly influenced the contents of this first book
on the history of regeneration research. The present work focuses on
several major scientific and philosophical issues that influenced and
were influenced by discoveries in animal regeneration. The following
paragraphs illustrate some of the issues whose histories are pursued in
subsequent chapters.

Lizard tail regeneration was demonstrated to the Paris Academy
of Sciences in the summer of 1686. The live presentations by Melchisedech Thevenot were followed, two years later, by a dissertation
on the general topic from Claude Perrault. A decade later, the same
body heard an anecdotal discussion of the regeneration of a human

1

fingernail after amputation of the finger tip (Jacques Roger, 1963, pp. 390–1; see also Chapter 6 in the present volume). But because the new spirit of science was just expanding its domain to include the experimental challenge of living nature, there was no experimental tradition upon which to elaborate these new observations, which in many ways linked generation and regeneration. The continuous amassing of curiosities of nature awaited only the advent of a new paradigm to initiate more fruitful inquiries into these wonderful phenomena.

In another more elaborate, though still largely descriptive, study of animal regeneration, R. A. F. de Réaumur reported his observations on crayfish appendage replacement to the academy in 1712 (discussed in Chapter 3 of this volume by Skinner and Cook); but it was Abraham Trembley's discovery of the extraordinary regenerative capacity of the polyp, reported to the academy by Réaumur and transmitted throughout the community of European natural historians, that gave rise to the modern discipline of regeneration research and to the larger field of experimental zoology (discussed in Chapter 4 by Lenhoff and Lenhoff).

The subsequent history of regeneration research is a window onto a scientific enterprise that continues to bear fruit. Indeed, it provides an opportunity to examine and analyze a life science that has helped to transform and has made important contributions to contemporary biology. Regenerative phenomena have been identified, defined, and explored at the organismic level in a vast array of species for more than three hundred years. The study of planarian regeneration is one of the most widely used "experiments" by which children are introduced to the wonders of nature. Nevertheless, we have only recently arrived at a stage of technological advancement that permits the resolution of fundamental questions about the regulatory mechanisms of generation and regeneration. Regeneration research has thus reached a turning point at which addressing the molecular basis of pattern regulation is experimentally feasible. The limits to the validity and explanatory power of theoretical constructs such as Wolpert's positional information model (Wolpert, 1969; see also Chapter 12 of the present volume) can now be tested more fully. As the excitement and potential of molecular biology attract more and earlier attention from students of developmental biology, it becomes increasingly important for those already engaged in exploring esoteric detail in regenerative phenomena, as well as those just entering the lists, to have ready access to the overall history behind, and whole-animal context of, the cellular and molecular relationships being elucidated.

In the organization of this book, historical context and scientific content have been interwoven to create a valuable instrument for nonspecialist and specialist, student and seasoned investigator. It contains scholarly explorations of the foundations of regeneration research, as well as of the philosophical issues that informed the interpretations of the field at specific periods in its development. In the final analysis, the history of regeneration research is also the history of the development and application of the scientific (i.e., experimental) method in the biological sciences.

The milestones selected for inclusion and analysis in this book are encountered in chronological sequence, for the most part. Following Goss's general survey and definition of the field (Chapter 2), the eighteenth-century research is presented in four essays. Three focus on the lives and contributions of naturalists (Skinner and Cook, on Réaumur; Lenhoff and Lenhoff, on Trembley; and Dinsmore, on Spallanzani) whose roles in the establishment of regeneration research were unquestionably of major importance. They framed many of the questions that are fundamental to present-day developmental biology, as the individual analyses reveal. The fourth treatise (Benson's essay, Chapter 6) is both analysis and synthesis, an incisive intellectual dissection and reconstruction of the context that framed the efforts of our eighteenth-century protagonists.

The evolution of regeneration research in the nineteenth century is presented somewhat differently, owing in large part to its eclipse by other events of great moment in the history of the life sciences. One is the unfolding of the drama that validated evolutionary biology. A case can be made, however, that the discovery of nerve-dependent regeneration, the neurotrophic phenomenon described by T. J. Todd in 1823, maintained the light transiently. It most certainly provided a point of departure for many of the exciting nerve and nerve-dependent regeneration studies of the present century as analyzed by Singer and Géraudie in Chapter 7. At any rate, a thread of continuity is revealed that shows the field to have gone from a relatively quiescent state, in most of the early decades of the century, to bloom in the final decades, as seen in Churchill's analysis in Chapter 8. Maienschein's essay (Chapter 9) then shows regeneration research erupting into the twentieth century and onto the American scene with the experiments and deliberations of T. H. Morgan, before he unceremoniously dropped his pursuit of regeneration in favor of his beloved fruitflies. It is ironic that the organism Morgan pursued in his escape from regeneration should have become in recent decades such an important model, via imaginal disc manipulation, for the investigation of pattern regulation in regenerating systems.

Although it is too early for a definitive analysis of the twentieth century, three areas of contemporary interest, represented in the final chapters, may turn out to have general significance in the evolution of regeneration research. Of these, one is a renaissance of investigations into bioelectrical phenomena, the early history of which often reads like a story in a tabloid newspaper. As Vanable's discussion in Chapter 10 reveals, this lurid early history has been a handicap to legitimate contemporary research in this field. Another area of continuing debate concerns the cellular origin of the blastema. This issue goes to the heart of questions about the differentiated state of cells and tissues and its stability. Only more recent technological advances, beginning with the development and perfection of cell and tissue culture techniques, have permitted approach to these historically debated issues, described in Liversage's essay (Chapter 11). And finally, one of the most exciting and productive, as well as provocative, areas of regeneration research has been and continues to be the development of explanatory theories for the varied expressions of regenerative responses. The history behind the rise of contemporary theories of positional information and pattern regulation, the false starts and partial successes that provide a framework for contemporary debates (food for future historians), is at the core of Wolpert's engaging critical closing chapter.

Two general themes will be readily discerned in the essays in this collection. First, it is clear that we have come a long way in our understanding of the natural world. From the data gathered by the authors of this volume, one readily perceives that regeneration research is more than just a footnote in the history of embryology. Moreover, one might even argue that the history of regeneration research offers an excellent example, a paradigm with which to understand the evolution of the life sciences. Second, and significantly for the general appreciation of the how and why of a science, these essays offer an occasional glimpse of the behind-the-scenes human factors that have a substantial impact on the dynamics of discovery and progress in any discipline.

This book does not presume to be either a comprehensive or definitive treatment of the history of regeneration research – merely a sampler to whet the appetite of those whose curiosity about the field has led them to seek a deeper understanding of its foundations. An obvious limitation is that only topics related to animal regeneration are discussed; regeneration in plants will have to await a subsequent volume. Moreover, exclusions resulting from the constraints of space and my own current understanding of the field are certainly open to criticism. It is hoped that future projects will address the gaps that appear to be most in need of filling. In some cases, the issues raised in the various chapters will challenge the reader to undertake a broader exploration. Everyone will find provocative observations simply noted, but not pursued, if they are not germane to the author's immediate objectives. These can be used to advantage by educators as a means of stimulating students to question assumptions made by any of the contributors. For example, the hematogenic origin of the regeneration blastema was postulated in the late nineteenth century to account for the source of the cells that make up the regenerate. (See Chapter 12.) Current interest in, and discoveries related to, growth factors in the immune system, their regulation of white blood cell activities, and their putative significance for regeneration (e.g., Sicard and Lombard, 1990) may demand a historical reevaluation of observations upon which the hematogenic model was based.

Finally, we have opened a new avenue of communication between scientists and historians and have, at the same time, begun exploring through a heretofore unused window onto the historical establishment of experimental biology. As the reader will readily discover, especially in the chapters covering the eighteenth and nineteenth centuries, the history of embryology is much more closely linked with, and indebted to, discoveries in regeneration than previous studies have allowed.

REFERENCES

Goss, R. J. 1969. *Principles of Regeneration.* Academic Press, New York.
Roger, Jacques. 1963. *Les sciences de la vie dans la pensée française du XVIIIe siècle.* Colin, Paris.
Sicard, R. E., and M. F. Lombard. 1990. Putative immunological influence upon amphibian forelimb regeneration. *Biol. Bull.* 178: 21–4.
Wolpert, L. 1969. Positional information and the spatial pattern of cellular differentiation. *J. Theoret. Biol.* 25: 1–47.

2

The natural history (and mystery) of regeneration

RICHARD J. GOSS

Introduction

Webster's Ninth New Collegiate Dictionary defines "natural history" as "the study of natural objects esp. in the field from an amateur or popular point of view." In the present context, I should like to redefine it to mean two things: history and nature. Accordingly, I shall first attempt to trace the evolution of humanity's way of approaching the fascinating problem of regeneration, from Paleolithic times through the classical period in Greece and into the Middle Ages. Since I do not pretend to be a historian, I suspect that this may fulfill the "amateur" aspect of the dictionary definition. I shall then elaborate on some of the unanswered questions about regeneration, questions that have tantalized and frustrated our predecessors for well over two centuries and that remain unanswered today. Finally, I shall attempt to define and classify *epimorphic* regeneration with respect to other developmental phenomena often included beneath the umbrella of the word "regeneration." So much for the nature of regeneration, which is more than enough to represent the popular point of view.

Myths and mysticism

One cannot investigate the immediate causes of regeneration without wondering about the ultimate ones. The latter are of two kinds. There is the more familiar question of how the capacity (or incapacity) to replace lost parts evolved in the first place. And then, there is the historical evolution of our awareness of such problems. This latter is an important dimension to research, because answers cannot be found until the appropriate questions have been asked. The appropriateness of questions depends on the perspective from which they are formulated. In scientific research, questions take the form of experiments. They must be defined and refined to the point where they back nature

7

into a tight corner, forcing it to betray its innermost secrets. Like the more rhetorical questions they resemble, experiments that are vaguely conceived yield the ambiguous answers they deserve. Only those that are precisely designed give definitive results.

The precision with which one formulates questions is a function of one's insight into a problem, which is in turn profoundly influenced by the historical, and even the philosophical, perspective inherited from our predecessors. (See Chapter 6, this volume.) It is useful, therefore, to explore what our ancestors thought about regeneration in order to trace the origins of our present, still evolving, ideas.

Paleolithic art in French and Spanish caves yields the earliest evidence that prehistoric people knew that they did not regenerate lost appendages. Using their own hands as stencils, they made negative silhouettes by spray painting around them. Such paintings have been found on the walls of caves in Castillo, Altamira, Pech Merle, Lascaux, and Gargas. In some cases, parts of fingers are missing (Giedion, 1962; Ruspoli, 1986). Whether this was the result of accident (e.g., frostbite) or ritual amputation is not known (Kuhn, 1955). Either way, one wonders if these early cave people harbored any expectations that their lost fingers might regenerate. If so, they were to be sadly disappointed.

Many millennia later, evidence for an awareness of regeneration was preserved in the mythology of ancient Greece. Here, allusions to regenerative events abound (Hamilton, 1942). Not the least of these mythological examples is the legend of the Hydra. Whether or not the Greeks were aware of the regenerative prowess of coelenterates, it is altogether fitting that one of our best-known examples of regeneration should have been named after the Greeks' fabulous creature, even though the latter could grow back two heads in place of one, whereas the hydra can only replace what is lost. It was the second labor of Hercules to slay the Hydra, a serpentlike creature purported to have nine heads (although a dozen or more often are depicted in illustrations). To inhibit the regeneration of the heads, Hercules had Iolaus cauterize the stumps with a firebrand, a technique that would have disinfected the wound and stopped the bleeding (Majno, 1975) but probably would not have prevented regeneration.

The legend of Prometheus is another case in point. He was punished by Zeus for having defied the gods' orders by giving the gift of fire to man, an act of insubordination for which we should all be grateful. As described by the epic poet Hesiod,

> . . . in fast bondage he bound Prometheus, the devious planner,
> whipping the painful bindings over a column at midpoint,

and against him sent a long-winged eagle to feed on his liver,
which was immortal; but whatever this long-winged bird ate
during the day grew during the night again to perfection.

(*Theogony*, lines 521–5)

One wonders if some unsung Greek biologist might have carried out
a partial hepatectomy on an unfortunate beast, presumably without
benefit of anesthesia. If so, the victim would have had to live long
enough to undergo liver regeneration. Whoever might have performed this feat of survival surgery apparently never published the
results – except in the annals of Greek mythology. We now know that
the liver does have this extraordinary capacity, and it is used routinely
as an experimental model for studying fundamental questions of
regeneration.

Regeneration of the eye was presaged in the tale of the three hags
in the legend of Mercury. According to this story, the hags had only
one eyeball among them, which they passed back and forth, each in
turn inserting it into her orbit to have a look around. Perhaps the
Greeks knew that scallops could regrow the eyes that adorn the mar-
gins of their shells or that certain amphibians can regenerate the lens
and retina. Indeed, the experiments by Stone (1963) in which he
removed entire eyeballs from newts, grafted them back into the orbit,
and noted visual recovery, are reminiscent of the eye-swapping practices of the three hags.

In each of these cases, the myth exaggerates reality – which is what
myths are supposed to do, of course. The regeneration of the Hydra's
head is an example of hypermorphic regeneration. In the legend of
Prometheus, the liver regenerated overnight, by which time a real
mutilated liver would not yet have even commenced DNA synthesis.
The fact that a liver can regenerate, however, has been confirmed by
Ingle and Baker (1957), who performed partial hepatectomies on
rats at monthly intervals for a year, showing that in each episode the
remnant could expand to its own original mass. Human livers also
regenerate, as they do after transplantation of hepatic fragments. The
instantaneous recovery of vision by the three hags reveals again that
the Greeks were unconcerned with the fact that growth and regeneration take time. Still another exaggeration – this one truly incred-
ible – is the story of how Cadmus planted dragon's teeth, which then
germinated into an army of soldiers. Not even Kollar and Fisher
(1980), who succeeded in inducing hens to grow teeth, have dupli-
cated this developmental tour de force.

In the Middle Ages, the prospects for epimorphic regeneration in
human beings was the subject of much speculation. Whether or not

medieval people were familiar with the regeneration of body parts in lower forms is not known. They were, of course, well aware that amputated human limbs do not grow back, an awareness that spawned no small measure of wishful thinking. As Price (1976) and Price and Twombly (1978) have suggested, the hope of regaining a missing arm or leg may have been inspired by the phantom limb phenomenon, in which an amputee can still feel the lost limb, often in great detail.

This real illusion, long known in folklore, was first formally described by the great French sixteenth-century barber-surgeon Ambroise Paré (1517–90). His skill as a surgeon and his experience with battlefield amputations exposed him to many amputees who related to him the phantom sensations, and sometimes pain, felt in their missing limbs. Though not well understood, the phantom limb phenomenon has attracted considerable clinical attention in recent years (Ewalt, Randall, and Morris, 1947; Simmel, 1958; Carlen, Wall, Nadvorna, and Steinbach, 1978; Siegfried and Zimmerman, 1981). In general, the feelings tend to focus on the ends of the limbs – the toes and feet or the fingers and hands. Over time, the perceived size of the phantom limb may grow smaller than the original. The ends of the limb may eventually come to be felt as being closer and closer to the stump, or sometimes even as being in the stump itself, a condition referred to as "telescoping." The phantom digits are often described in the flexed position. There are even records of patients who have undergone surgical removal of teeth, or even of a breast or penis, experiencing the feeling of still possessing these parts of their bodies. The lack of such feelings in infants or in the case of congenital absence of limbs may explain why there are no accounts of phantom foreskins following circumcision. Whether or not experimental animals feel phantom appendages (e.g., amphibian limbs or tails, or deer antlers) we shall never know.

The phantom limb phenomenon, now believed to result from a reaction of the central nervous system to the lack of nerve impulses from the severed limb (Wall, 1981), may once have created the impression that a limb's "essence" survived its amputation and that maybe there was hope of reversing the process. Admiral Nelson (1758–1805), who lost his right arm in the battle of Tenerife, in 1797, when a musket ball smashed into his elbow, is said to have believed that the phantom limb he later felt was proof that his soul would survive the disappearance of its material substrate.

Religious overtones were to be expected in people's wishes to make up for what nature lacked in prescientific times. So it was that certain "miracles" have been recorded in which, through divine intervention, amputees were said to be restored to their original form. These events, attributable in part to prevailing belief in the supernatural, as well as to the superstitious minds of people in the Middle Ages, were probably enhanced by "the pain that remains after amputation" (*dolar membri amputati*), as Price and Twombly suggest (1978). These authors have cited the following examples of purported miracles.

One story from the seventh or eighth century describes the miraculous reattachment of the amputated right hand of Saint John of Damascus. Unjustly accused of writing a treasonous letter to the caliph of the Saracens, he was punished by having his hand cut off. Taking the severed hand to a church and holding it against the stump from which it had been separated, he pleaded with the Mother of God to heal him. He then fell asleep, dreamed that the holy Mother had intervened on his behalf, and awoke to find his amputated hand once again intact and functional on his arm.

Early in the twelfth century, Peter of Grenoble lost a leg when he was struck by lightning. This was no simple accident, for it supposedly occurred when Peter continued his plowing despite his priest's prohibiting work on that day – the birthday of the blessed Mary Magdalene. Surviving his injuries, the crippled Peter devoted himself to the Church. One night, he had a vision that the Virgin Mother had ordered his leg to be restored, whereupon the parts of his missing leg were reassembled onto the stump and he became whole again.

Another miracle was that of Miguel Juan Pellicero, whose right leg had been amputated in the hospital at Zaragoza, Spain. After his stump had healed with scar tissue, he dreamed one night in 1640 that he anointed himself with oil from the lamps in the Chapel of Our Lady del Pilar and that she had promised to give back his leg. When he awoke, sure enough his leg had been restored, although it took a few days to get its feeling back.

Such accounts as these include nearly instantaneous replacement of lost members, often involving regrafting of severed parts, rather than their regeneration. Such transformations usually were said to have happened while the subject was having a vision or dreaming, which may be interpreted, metaphorically, as a cover-up for the lack of understanding of how such miraculous results could have been achieved. One may conclude that if the process of epimorphic regeneration in animals had been known in those days, it most certainly would have figured prominently in the fantasies and legends of the times. Instead, human imagination resorted to more mystical explanations of hoped-for miracles.

Even in scientific circles, the idea of the nonmaterial "essence" of a missing part has been referred to in recent years. The Kirlian effect (Johnson, 1975) is a case in point. If an object is placed in contact with a photographic emulsion on film, the developed film reveals exposed areas – presumably electrical in origin – at and around the points of contact. For example, when a leaf is tested, the image shows the outlines of the veins. In some cases in which part of the leaf had been cut off, however, the exposed film shows an outline of the missing part. This has been interpreted as evidence that the "essence" of the absent portion of the leaf creates the image, but careful attention to the methods used reveals a simpler explanation. The whole leaf is first placed on the film, and then a portion of it is cut away. Clearly, the image of the missing part is, in all likelihood, created by residual material from the surface of the leaf that was deposited on the film before the part was removed.

In regeneration studies, the concept of morphogenetic "fields" has helped us understand the developmental potential of amputation stumps. Such fields are merely abstractions of the prospective development of missing parts from an amputated stump during its regeneration. As Weiss (1939, p. 266) was careful to point out, however, this concept has nothing to do with "disembodied vital principles" but is merely a heuristically useful way to explore the morphogenetic capabilities of regenerating stumps.

We have come a long way from the primitive minds of Paleolithic cave artists, from the myths of ancient Greece, and from the superstitions of the Middle Ages; yet even today, is there any regenerationist who does not still hope that one day we shall find out how to promote mammalian regeneration? Oscar Schotté used to say, only half jokingly, that he would give his right arm to discover that secret. Yet even if we could induce the regrowth of arms and legs in human beings, the problem of regeneration would remain unsolved. It would remain unsolved because we still do not understand how it is that the morphogenesis of regeneration buds can give rise to such complex structures from seemingly unorganized masses of cells.

Unsolved problems of regeneration

It has been more than two hundred years since regeneration of animal parts was discovered by the eighteenth-century naturalists whose work is described in subsequent chapters of this volume. From the very beginning, the subject has excited wonder and curiosity. In 1958, Newth described some of the early reactions:

In 1768 the snails of France suffered an unprecedented assault. They were decapitated in their thousands by naturalists and others to find out whether or not it was true, as the Italian Spallanzani had recently claimed, that they would then equip themselves with new heads. . . . Thus, the study of regeneration in animals . . . became, perhaps, the first of all branches of experimental biology to be popularized.

Among the successful experimenters was Voltaire, who . . . marveled briefly; saw at once that the loss and replacement of one's head presented serious problems for those who saw that structure as the seat of a unique "spirit" or soul; and thought of the possible consequences of the experiment for man. . . . Later, he expressed confidence that men would one day so master the process of regeneration that they too would be able to replace their heads entire. There were many people, he implied, for whom the change could hardly be for the worse. (pp. 47–8)

The idea of surviving decapitation was not without precedent, especially in France. Saint Denis (or Saint Dionysius), who became the first bishop of Paris in the third century, was beheaded on October 9, 250 C.E., whereupon he is said to have walked from Montmartre to St. Denis, carrying his own head. Accordingly, the future patron saint of France was also declared patron saint against headaches (Holweck, 1924).

Although the uneven distribution of regenerative abilities in the animal kingdom was undoubtedly noted in pre-Darwinian times, the theory of evolution lent an entirely new perspective to the problem. For the first time it was possible to contemplate the evolution of a developmental process resulting in regeneration. This raised questions about the adaptive nature of regeneration and whether it was monophyletic or polyphyletic – that is, whether it evolved once or many times. Today we are still seeking answers to this problem (Reichman, 1984).

With respect to the phylogeny, or evolutionary history, of regeneration, we must determine whether regenerative abilities have been selected for or against in the course of vertebrate evolution. If there has been selection for, then each example of epimorphic regeneration must have arisen separately in this or that appendage, presumably as an adaptation to environmental pressures. In that case, one might assume that each appendage evolved its own mechanism of regrowth. There are indeed differences in the developmental mechanisms of regeneration in various appendages, but these seem only to reflect the innate differences in anatomical structure that distinguish one appendage from another. Tails possess spinal cords that other appendages lack. Fins and barbels are essentially without muscles, which eliminates the need for myogenesis that one sees in limbs and tails.

Nevertheless, each of these structures goes about the process of replacing itself in much the same way. All of them undergo epidermal wound healing, loss of cell specializations (dedifferentiation), formation of a regeneration bud or blastema made of dedifferentiated cells, and development that starts at the level of the stump and proceeds outward (proximodistal morphogenesis). They all even depend on the nervous system to form blastemas. It would seem too coincidental if every regenerating appendage invented the very same sequence of developmental processes independently of the others. Perhaps there is only a single way in which structures can regenerate themselves after amputation, but that seems unlikely.

One can imagine other ways to achieve the same ends. For example, each tissue in the stump might have regrown itself from its own cells, rather than mixed its dedifferentiated cells into an aggregate formed with cells from neighboring tissues. Even if a blastema is necessary, is there any reason why it should differentiate in a proximodistal direction, rather than the other way around? Either way, one would expect the same outcome. Hypothetically, there could be more than one way to replace an amputated appendage; yet they all do it in much the same manner.

There may be two reasons for this. One is that different structures regenerate the way they do because regeneration "recapitulates ontogeny." Regenerationists have long recognized the similarities between blastemas and the limb buds or tail buds of embryos (Stocum and Fallon, 1984). Except perhaps for amputation and dedifferentiation, which are either impossible or unnecessary in an embryo, for obvious reasons, the way an appendage develops prenatally has much in common with how it regrows (if it can) postnatally. The similarity between embryogenesis and regeneration is witnessed in the similarity of their final products, as Needham has emphasized (1961). This reflects the fact that development is a highly conservative process, employing the same mechanisms (with minor variations) again and again in the process of maturation.

Jacob (1977) pointed out that evolution (including the evolution of development) has occurred more by tinkering than by engineering. The same basic processes have been used over and over again, not because this is more economical than synthesizing a whole new set of genes every time there is an evolutionary advance to be made, but because it is the only way the animal knows. Organisms are in fact incapable of inventing new genes from scratch. Rather, they tinker with those they have, as their genes mutate in a haphazard way, until a change occurs that happens to be an adaptational improvement.

Thus, developmental mechanisms are prisoners of their own phylogenetic histories, carrying with them the baggage of their evolutionary
pasts. In regeneration, an appendage's embryonic heritage is reflected
in the way it regrows. Perhaps the only mechanism available is the one
that was there, namely, the genetic instructions by which the embryo
sprouts its original complement of appendages. It may have made
sense to reuse the preexisting program rather than to compose a new
one every time something needed to be regenerated.

The second reason for the similarity between different examples of
epimorphic regeneration could be that this ability is such a primary
attribute of animals that the potential is universally distributed
(Korschelt, 1927). The spotty distribution of its expression could be
the result of its having been selected against, rather than for, in some
animals. When its occurrence would have been detrimental or impossible, for any reason, or even when it might have been neutral with
respect to the survival and perpetuation of its possessors, it would
have been eliminated by natural selection. In this scenario, one
expects examples of regeneration always to resemble one another
closely, as they do, because they are derived from that same ubiquitous
process that transcends all vertebrates.

In answer to the question of whether regeneration is monophyletic
or polyphyletic, it would appear to be some of both. It is monophyletic in the sense that the mechanisms by which it occurs evolved once,
as did their antecedent embryonic processes. Yet it is polyphyletic in
the sense that each time regeneration has been eliminated it has happened separately. Thus, the absence of regeneration is polyphyletic.
The question that remains to be addressed is whether the abolition
of regeneration has been achieved in the same way each time it has
occurred.

Unhappily, we know little about why nonregenerative appendages
do not regenerate. There are two ways to approach this problem, the
proximate and the ultimate. The proximate explanation relates to
the physiological and developmental mechanisms responsible for the
failure of an amputated structure to regenerate. These are of two
kinds: errors of omission and errors of commission. Conceivably,
regeneration can be subverted by omission of one or more essential
ingredients. Experiments with denervation in frogs suggest one possible factor in the extinction of regeneration. The fact that regeneration can be enhanced in postmetamorphic frogs by augmenting the
nerve supply (Singer, 1951, and Singer and Géraudie, Chapter 7, this
volume) suggests that the capacity to regenerate may have been lost
in some animals because of the depletion of nerves below a necessary

threshold (Rzehak and Singer, 1966). The inability to promote regeneration in mammals in this way, however, may mean that other factors are involved.

Alternatively, the capacity to regenerate may lie latent in all nonregenerative appendages, precluded only by the existence of a blockade, either physiological or anatomical. It is commonly supposed that the premature development of a dermal scar between the wound epidermis and underlying mesoderm prevents blastema formation. Whatever the explanation is for one system, it could be something entirely different for others. Errors of commission may come in many forms.

The ultimate explanation for the heterogeneous distribution of regenerative abilities in the animal kingdom has traditionally focused on the adaptive nature of regeneration. In animals that frequently lose appendages, the ability to grow back the missing part would be a selective advantage. Thus, liability to injury is a self-fulfilling prophecy. This interpretation, if true, would make sense of the bewildering occurrence of regeneration in some animals but not in others. "At the present time," wrote Voronsova and Liosner (1960, p. 408), "not a single investigator would be willing to predict the nature of regeneration in a species of unknown regenerative powers from its known position in the taxonomic scale." Liability to injury was a concept of considerable appeal for early regenerationists, as Morgan's review reveals (1901), but Morgan himself argued vehemently against the idea that regeneration could have arisen by the natural selection of those appendages in greatest jeopardy. (See Chapter 9 of the present volume.) Noting that the third, fourth, and fifth legs of the hermit crab, for example, are in a highly protected location, yet can regenerate following amputation, he was at a loss to understand how their replacement could have had a selective advantage. He pointed out that among animals in general, those individuals that successfully avoid appendage loss would have greater selective advantages than those that do not. The implication was that regeneration might more logically have been selected against in those cases in which it was unnecessary. This in itself, however, is also a form of adaptation.

Needham (1961) disputed Morgan's data and conclusions about hermit crab regeneration, showing not only that the "protected" legs can be lost in nature but also that they can play a role in natural selection. "Morgan tended to minimize quantitative differences in ability," Needham wrote, "and to magnify differences in susceptibility to loss."

In response to Needham's defense of the adaptive nature of regeneration, Spilsbury (1961) questioned how groups of genes different from those active in embryogenesis could be responsible for the regeneration of the same end product by natural selection. Such convergence, or "equifinality," though not always perfect, is more logically explained by the theory that the same genes are switched on for both embryonic development and regeneration. Arguing against Spilsbury's view, Barr (1964) maintained "that the occurrence of regeneration can be thought of as the result of a break-down in that delicate equilibrium which constitutes the postembryonic organism and that it requires no further genetic information beyond that which determined the organism's embryonic development." Presumably, Barr's "break-down in that delicate equilibrium" may itself depend on appropriate mutations subject to natural selection.

Classification of regeneration

Transcending all discussions of regeneration is the difficult problem of defining our terms. The word "regeneration" is a generic term applied to a wide variety of phenomena, but we are discussing "epimorphic regeneration." Epimorphic regeneration refers to the regrowth of amputated structures from an anatomically complex stump. This may involve either the replacement of parts of appendages or the regeneration of fractions of organisms into new complete individuals after their bisection. Whatever it is that must be regenerated, the sequence of events by which its replacement is achieved is essentially the same. The first event in epimorphic regeneration is the development of a blastema, or regeneration bud, derived from dedifferentiated cells, out of which the new structure will take shape. This kind of regeneration, then, is a specific developmental phenomenon, yet not without its ambiguities.

Epimorphic regeneration is unique in the spectrum of developmental events in animals because it is not universally distributed. All metazoans undergo embryogenesis, and presumably aging. All of their cells and tissues are subject to constant turnover, or physiological regeneration. All tissues, with the exception of teeth, exhibit wound-healing responses (tissue regeneration) after interruption of their continuity. And practically all organs of the body are capable of compensatory growth when overworked, as when the liver regenerates after its partial removal.

All of these developmental events except epimorphic regeneration are ubiquitous among animals. The only other process that is unevenly distributed is metamorphosis, such as what occurs in the transformation of a caterpillar into a butterfly or a tadpole into a frog. These differ from regeneration, however, in that all individuals capable of metamorphosis undergo it. Nevertheless, these two phenomena have much in common, including the ability to give rise to new structures in postembryonic individuals.

Of particular interest is the controversial relationship between epimorphic regeneration and tissue regeneration (Carlson, 1970, 1978; Goss, 1984; Korneluk and Liversage, 1984). Are they to be regarded as expressions of the same basic phenomenon, differing only in degree, or should they be interpreted as qualitatively distinct developmental events? One approach to resolving this problem is to examine their similarities and differences.

There are, of course, many similarities between epimorphic regeneration and the mechanisms by which the various tissues and organs of the body repair themselves after injury. In both cases, restorative processes are triggered by local wounding. Both depend on the survival of a remnant of the original structure as a source of cells for new development. This development is achieved by similar cellular processes, including migration, aggregation, proliferation, and differentiation. Both end products resemble the original structures that have been replaced, although imperfect replicas are not uncommon. In each kind of regeneration, the mechanism by which it is achieved appears to recapitulate the original ontogeny. Thus, there are many ways in which these two modes of regeneration are alike. There are more, however, in which they differ.

First of all, although epimorphic regeneration is sporadically distributed in the animal kingdom, tissue regeneration is almost universally represented, for virtually all tissues are capable of repairing themselves (McMinn, 1969; Goss, 1978). It must be concluded that the regeneration of individual tissues is more important than the replacement of lost appendages in the overall economy of nature.

The actual mechanisms of these two kinds of regeneration are also different. Although both are responses to injury, epimorphic regeneration cannot proceed without epidermal wound healing. Tissue regeneration does not require the growth of wound epidermis (except for the healing of integumental wounds). If no wound epidermis develops after amputation, no blastema forms, but internal tissues can nonetheless repair their injuries.

In the case of epimorphic regeneration, extensive cellular dedifferentiation is an indispensable prelude to blastema formation. This dedifferentiation is so great as to allow for the possibility of metaplasia in subsequent stages of development. In tissue repair the cells may also lose some of their specializations, but not, apparently, to the extent that they can redifferentiate into histologically unrelated tissues. Tissue repair may involve the temporary modulation of participating cells, but as far as we know these cells retain their basic identities.

By means of migration and proliferation, the cells in the vicinity of an injury converge. In amputation stumps, they accumulate beneath the wound epidermis to give rise to a blastema. In internal tissues, they aggregate between the severed components whose continuities have been interrupted. Such repair aggregates are variously referred to, for example, as "granulation tissue" in the case of the dermis or as a "fracture callus" between the ends of a broken bone. Comparable clusters of cells are associated with other tissues: fibroblasts with severed tendons, satellite cells and myoblasts with muscles, and ex-Schwann cells with nerves. Despite the superficial similarity between repair aggregates and blastemas, they appear to be different structures. Blastemas, by definition, must be enveloped in epidermis and cannot fulfill their developmental potential without it. Repair aggregates can continue to develop without direct communication with epidermis.

The process of morphogenesis also differs in these two types of regeneration. In epimorphosis, the blastema differentiates in a proximodistal direction (toward the end of the appendage), perhaps reflecting the comparable polarity seen in developing limb buds or tail buds in embryos. In tissue regeneration, differentiation takes place either simultaneously, throughout the repair aggregate, or in a centripetal direction, as the peripheral cells differentiate before the central ones. The latter polarity could be attributed to, if not dependent on, the sequence of revascularization in the traumatized region.

In some appendages, adequate innervation is necessary for epimorphic regeneration to occur (Singer, 1960, and Chapter 7 of the present volume). Without sufficient nerves, no blastema is formed. On the other hand, neurotrophic influences are unnecessary for tissue repair. Fractured bones and wounded skin heal themselves normally in denervated appendages. Even injured skeletal muscle undergoes repair in the absence of innervation, although the muscle fibers may eventually atrophy.

Perhaps the most important difference between epimorphic regeneration and tissue regeneration relates to the extent to which lost parts are replaced. In epimorphic regeneration, not only is the continuity of severed tissues in the stump completed, but entirely new structures are formed, even in the absence of their counterparts in the stump. As Spallanzani noted more than two hundred years ago (see Chapter 1 of the present volume), new skeletal elements differentiate. Entire muscles are formed. Limbs often reconstitute their joints and digits, and tails give rise to individual missing segments. Morphogenesis is sufficiently extensive in most regenerating appendages to replace the amputated parts entirely. In contrast, the degree of repair in tissue regeneration suffers by comparison. It is essentially a stopgap measure whereby the interrupted continuities of the injured tissues are completed. New tissue may be formed in the process, but this is not enough to give rise to additional structures. Thus, these two modes of repair are well adapted to the different tasks they are called upon to perform. Because the degrees of injury to which they respond are so widely divergent, it is not surprising that each of these processes is a distinct developmental phenomenon. Carlson (1970, 1978) has pointed out that these two kinds of regeneration can take place in one and the same newt limb. Following amputation, therefore, they may occur simultaneously, but it is epimorphic regeneration that dominates (Dinsmore, 1974).

These arguments make it quite plausible that epimorphosis and tissue regeneration are qualitatively different events that have evolved to correct separate problems. Nature, however, does not always conform to the categories defined by biologists. There are certain hard-to-define phenomena that defy our best attempts to assign them to discrete compartments. Regeneration is no exception.

It is well known, for example, that bones cannot regenerate following their complete removal, presumably because there is nothing left behind from which to derive the necessary cells. In young rats, however, it has been shown that excision of the expanded head of a bone (the epiphysis and its associated cartilaginous plate) may result in its complete replacement (Nunnemacher, 1939; Libbin and Weinstein, 1986). This phenomenon, if substantiated, resembles epimorphic regeneration in that the new structure appears to be an outgrowth of a stump in a longitudinal orientation. It seems to be intermediate between fracture healing and epimorphic regeneration.

The regulation of embryonic development is another case in point. Experimental resection of body parts in amphibian embryos, for example, may be followed by normal development, not through the establishment of a blastema but by reorganization of remaining cells, a phenomenon akin to morphallaxis, the process by which a "part is transformed directly into . . . part of an organism without proliferation at the cut-surfaces" (Morgan, 1901, p. 23).

Lens regeneration in the amphibian eye, that remarkable process whereby a missing lens is replaced by a new one derived from the dorsal iris, is achieved by metaplasia but involves neither epidermal wound healing nor blastema formation. (See Chapter 8 of the present volume.) It is more than mere tissue repair, yet it is not epimorphic. Where does it fall in the spectrum of developmental phenomena? Clearly in a class by itself.

Korneluk and Liversage (1984) have compared the regrowth of amputated legs in *Xenopus* froglets with the regeneration of newt limbs. In the former, only an unbranched, tapered extension of the limb stump develops, consisting mostly of cartilage with sparse muscle confined to the portion near the body. Noting a distinct paucity of dedifferentiation in *Xenopus* and a difficient "fibroblastema," they were inclined to classify these outgrowths as an example of tissue regeneration rather than epimorphic regeneration. Inasmuch as epimorphic regeneration depends on wound epidermis whereas tissue regeneration does not, the true nature of *Xenopus* leg regrowth could be resolved by inserting skinless stumps into internal locations where no epidermal wound healing could occur. If they still regrew, it would be tissue regeneration. If not, it would be epimorphic.

Similar questions could be raised about certain cases of mammalian regeneration (Goss, 1980, 1984). The ingrowth of marginal tissue from around full-thickness holes cut through rabbit ears may be difficult to distinguish from more conventional examples of appendage regeneration, except for the lack of proximodistal differentiation. The annual replacement of deer antlers, which appears superficially to be a true case of epimorphic regeneration, could be interpreted as an exaggerated version of scar formation. Even the interesting examples of finger tip regeneration in children could represent the summation of individual tissue repairs, especially bone overgrowth (Aitken, 1963; Speer, 1981), rather than regrowth of the epimorphic kind.

It is sometimes disconcerting to find that certain phenomena of regrowth do not fit neatly into our definitions of epimorphic or tissue regeneration. This may be in part our own fault for not making our definitions sufficiently exclusive. It is also in part the result of nature. A wide spectrum of reparative processes have evolved in animals, not all of which fall within the artificial boundaries of our definitions. As frustrating as this may sometimes be, it at least provides experimental biologists with a rich diversity of regenerative phenomena on which to test their hypotheses about the basic nature of development. Such is the history of the scientific enterprise.

REFERENCES

Aitken, G. T. 1963. Surgical amputation in children. *J. Bone Joint Surg.* 45A: 1735–41.

Barr, H. J. 1964. Regeneration and natural selection. *Am. Nat.* 98: 183–6.

Carlen, P. L.; P. D. Wall; H. Nadvorna; and T. Steinbach. 1978. Phantom limbs and related phenomena in recent traumatic amputations. *Neurol.* 28: 211–17.

Carlson, B. M. 1970. Relationship between the tissue and epimorphic regeneration of muscles. *Am. Zool.* 10: 175–86.

  1978. Types of morphogenetic phenomena in vertebrate regenerating systems. *Am. Zool.* 18: 869–82.

Dinsmore, C. E. 1974. Morphogenetic interactions between minced limb muscle and transplanted blastemas in the axolotl. *J. Exp. Zool.* 187: 223–32.

Ewalt, J. R.; G. C. Randall; and H. Morris. 1947. The phantom limb. *Psychosom. Med.* 9: 118–23.

Giedion, S. 1962. *The Eternal Present: The Beginnings of Art.* Pantheon, New York.

Goss, R. J. 1978. *The Physiology of Growth.* Academic Press, New York.

  1980. Prospects for regeneration in man. *Clin. Orthop.* 151: 270–82.

  1984. Epimorphic regeneration in mammals. In *Soft and Hard Tissue Repair* (T. K. Hunt, R. B. Heppenstall, E. Pines, and D. Rovee, eds.), pp. 554–73. Praeger, New York.

Hamilton, E. 1942. *Mythology.* Little, Brown, Boston.

Hesiod. *Theogony.* In *The Poems of Hesiod* (R. M. Frazer, trans.). University of Oklahoma Press, Norman. 1983.

Holweck, F. G. 1924. *A Biographical Dictionary of the Saints.* Herder, St. Louis.

Ingle, D. J., and B. L. Baker. 1957. Histology and regenerative capacity of liver following multiple partial hepatectomies. *Proc. Soc. Exp. Biol. Med.* 95: 813–15.

Jacob, F. 1977. Evolution and tinkering. *Science* 196: 1161–6.

Johnson, K. 1975. *The Living Aura: Radiation Field Photography and the Kirlian Effect.* Hawthorne, New York.

Kollar, E., and C. Fisher. 1980. Tooth inductions in chick epithelium: Expression of quiescent genes for enamel synthesis. *Science* 207: 993–5.

Korneluk, R. G., and R. A. Liversage. 1984. Tissue regeneration in the amputated forelimb of *Xenopus laevis* froglets. *Can. J. Zool.* 62: 2283–2391.

Korschelt, E. 1927. *Regeneration und Transplantation,* vol. 1. Borntraeger, Berlin.

Kuhn, H. 1955. *On the Track of Prehistoric Man.* Random House, New York.

Libbin, R. M., and M. Weinstein. 1986. Regeneration of growth plates in the long bones of the neonatal rat hindlimb. *Am. J. Anat.* 117: 369–83.

McMinn, R. M. H. 1969. *Tissue Repair.* Academic Press, New York.

Majno, G. 1975. *The Healing Hand: Man and Wound in the Ancient World.* Harvard University Press, Cambridge, Mass.

Morgan, T. H. 1901. *Regeneration.* Macmillan, New York.

Needham, A. E. 1961. Adaptive value of regenerative ability. *Nature* 191: 720–1.

Newth, D. R. 1958. New (and better?) parts for old. *New Biology* 26: 47–62.

Nunnemacher, R. F. 1939. Experimental studies on the cartilage plates in the long bones of the rat. *Am. J. Anat.* 65: 253–89.

Price, D. B. 1976. Miraculous restoration of lost body parts: Relationship to the phantom limb phenomenon and to limb-burial superstitions and practices. In *American Folk Medicine: A Symposium* (W. D. Hand, ed.), pp. 49–71. University of California Press, Berkeley and Los Angeles.

Price, D. B., and N. J. Twombly. 1978. *The Phantom Limb Phenomenon: A Medical, Folkloric, and Historical Study.* Georgetown University Press, Washington, D.C.

Reichman, O. J. 1984. Evolution of regeneration capabilities. *Am. Nat.* 123: 752–63.

Ruspoli, M. 1986. *The Cave of Lascaux.* Abrams, New York.

Rzehak, K., and M. Singer. 1966. Limb regeneration and nerve fiber number in *Rana sylvatica* and *Xenopus laevis. J. Exp. Zool.* 162: 15–22.

Siegfried, J., and M. Zimmerman (eds.). 1981. *Phantom and Stump Pain.* Springer, Berlin.

Simmel, M. L. 1958. The conditions of occurrence of phantom limbs. *Proc. Am. Philos. Soc.* 102: 492–500.

Singer, M. 1951. Induction of regeneration of forelimb of the frog by augmentation of the nerve supply. *Proc. Soc. Exp. Biol. Med.* 76: 413–16.

   1960. Nervous mechanisms in the regeneration of body parts in vertebrates. In *Developing Cell Systems and Their Control* (D. Rudnick, ed.), pp. 115–33. Ronald Press, New York.

Speer, D. P. 1981. Pathogenesis of amputation stump overgrowth. *Clin. Orthop.* 159: 294–307.

Spilsbury, R. J. 1961. Evolution of regeneration, and its possible bearing on philosophy. *Nature* 192: 1254–5.

Stocum, D. L., and J. F. Fallon. 1984. Mechanisms of polarization and pattern formation in urodele limb ontogeny: A polarizing zone model. In *Pattern Formation: A Primer in Developmental Biology* (G. M. Malacinski, ed.), pp. 507–20. Macmillan, New York.

Stone, L. S. 1963. Vision in eyes of several species of adult newts transplanted to adult *Triturus v. viridescens. J. Exp. Zool.* 153: 57–68.

Vorontsova, M. A., and L. D. Liosner. 1960. *Asexual Propagation and Regeneration.* Pergamon, London.

Wall, P. D. 1981. On the origin of pain associated with amputation. In *Phantom and Stump Pain* (J. Siegfried and M. Zimmerman, eds.), pp. 2–14. Springer, Berlin.

Weiss, P. 1939. *Principles of Development.* Holt, New York.

3

New limbs for old: some highlights in the history of regeneration in Crustacea

DOROTHY M. SKINNER and JOHN S. COOK

At the beginning of the eighteenth century, people living near the sea or by riverbanks were familiar with regeneration in Crustacea, notably crabs, lobsters, and crayfish. This familiarity derived in part from the observation of smaller than normal-size limbs on animals in the wild. The earliest, and for many years the most substantial, systematic observations were those of René-Antoine Réaumur (Figure 3.1), who is most famous today for the temperature scale named after him in honor of his invention of the alcohol thermometer. In his landmark paper on regeneration in crayfish, which was presented to the French Academy in November 1712 (Réaumur, 1712), Réaumur begins by chastising those learned people, the *savants* (Réaumur's spelling was *Sçavans,* with a capital *S,* but then, he capitalized many of his nouns), who scoffed at regeneration as "folk tales" (*contes du peuple*). Réaumur then defends his position by using the modern approach of elegant experiments coupled with shrewd observation.

But before discussing these, we shall describe this remarkable man. He was characterized in the late nineteenth century by T. H. Huxley as follows: "From the time of Aristotle to the present day I know of but one man who has shown himself Mr. Darwin's equal in one field of research – and that is Réaumur" (cited in Gough, 1975, n. 5). Réaumur was born in 1683 in the seaport of La Rochelle, France, and died seventy-four years later after a riding accident in St. Julien de Terroux in the *département* of Maine. His father died when Réaumur was nineteen months old, and he was reared by his mother and a cluster of aunts and uncles. At sixteen he went with an uncle to Bourges, where

By acceptance of this article, the publisher or recipient acknowledges the U.S. Government's right to retain a nonexclusive, royalty-free license in and to any copyright covering the article. Research supported by grants from the National Science Foundation, and the Exploratory Studies Program of the Oak Ridge National Laboratory to D. M. S. and by the Office of Health and Environmental Research, U.S. Department of Energy, under contract DE-ACO5-84OR21400 with Martin Marietta Energy Systems, Inc.

Figure 3.1. René-Antoine Ferchault de Réaumur, from an engraving by Simonneaux in the Bibliotèque Nationale, Paris. Courtesy of M. P. A. Lémoneaux (and the Archives of the Library of the Marine Biological Laboratory, Woods Hole, Mass.).

he studied law for three years, and then he moved to Paris to live with a cousin. His astonishing mathematical talents were discovered by an Academician in Paris, and as a consequence of his mathematical contributions he was nominated to the French Academy of Sciences as a "student geometer" at the age of twenty-five. In 1713, a short time after Louis XIV established the Paris Academy and when Réaumur was thirty, the new institution, rivaling the French Academy, was charged with generating an encyclopedia of all the arts, industries, and professions, and the job of writing it was given to Réaumur. Although he was neither a trained engineer nor a metallurgist, he was a keen observer and a person of considerable intelligence. He developed a deep interest in the iron, tin, and gold industries, and the "Réaumur process" for making steel would be used on a large scale a century later. He attempted to produce porcelain of the same quality as that found in excellent Chinese porcelains, unaware that the secret lay in a uniquely Chinese mineral called *petuntse*. Not surprisingly for one so interested in the properties of materials, he read about mollusk shells, and this reading brought him into the zoological sciences. The scope of the eighteenth-century "natural philosopher" is astonishing.

By modern standards, some of his and his contemporaries' ideas are best described as quaint. For example, they used the term "insect" for any small invertebrate and even for some large vertebrates that appear segmented. Réaumur said, "The crocodile is certainly a fierce insect but I am not in the least disturbed about calling it one" (Gough, 1975, p. 331). This curious remark reminds us to place his 1712 paper in context. In 1712 Linnaeus was only five years old. Robert Hooke, using the newly invented microscope, had described cells in cork fifty years earlier, and shortly thereafter Antonie van Leeuwenhoek himself had seen erythrocytes and spermatozoa, but it was to be a century and a half before Schleiden and Schwann put forward a general cell theory, before Virchow enunciated that "all cells come from cells," and before Pasteur put down the dogma of spontaneous generation.

Leeuwenhoek's visualization of spermatozoa was strong support for the idea that they were the homunculi from which humans (and animals in general) developed. That is, when "planted" in a favorable environment (i.e., women), these preformed tiny creatures simply enlarge. A natural corollary was the idea of preformation; thus a regenerating limb was thought by many (excepting, of course, those *savants* who did not believe in regeneration in the first place) as arising from an "egg" containing a miniscule preformed limb, ready to expand when needed, lying at the base of an amputated stump.

Réaumur's 1712 paper is on "reproduction" (by which he clearly meant regeneration) in lobsters, crabs, and crayfish. During a stay by the sea, to demonstrate the reality of regeneration he amputated legs of crabs, placed the crabs in containers with holes in the sides (the kind of containers used by fishermen to preserve live catches), and returned the containers to the sea. Alas, Réaumur's containers filled with sand, the experiment was without result, and he had to return to Paris. A similar experiment, carried out with freshwater crayfish, however, was a resounding success, and he was able to make detailed observations on limb regeneration.

Regeneration in Crustacea is inextricably linked to molting (reviewed in Skinner, 1985b), and to discuss regeneration we also need to describe some facts about molting. Like all arthropods, Crustacea grow by the periodic shedding of the old exoskeleton and the emergence of the animal within a new, and initially soft, exoskeleton, which then expands to encase the larger animal. After the expansion, the exoskeleton hardens to its characteristic texture. Dramatic developmental changes, including metamorphosis and puberty, occur only at the time of the appropriate molts in the animal's early life, but these developmental steps are not a concern here. Regeneration is clearly observed in molting adult Crustacea, uncomplicated by the major events associated with differentiation and sexual maturation.

Most of a crustacean's adult life is spent between molts in anecdysis (the intermolt period). During this period, two major interacting factors prevent growth (reviewed in Skinner, 1985a, b): the secretion by neurosecretory tissue (the X-organ–sinus gland complex, located in the eyestalk of higher crustaceans; Bliss and Welsh, 1952) of a molt-inhibiting hormone (MIH), and the lack of circulating, growth-stimulating steroids, or ecdysteroids. If one or a few limbs are lost during anecdysis, a small papilla forms at the base of the missing limb, but there is no further development during anecdysis. In due course and in response to suitable signals (some of them unknown), MIH secretion is stepped down, ecdysteroid levels rise, and the animal undergoes extensive preparation for ecdysis. These include, in some species, the development of calcareous gastroliths (calcium-storage structures at each corner of the gastric caeca, or stomach; Réaumur, 1712; Travis, 1960) and, in virtually all species, the resorption of most of the two innermost layers of the old exoskeleton (O'Brien, Kumari, and Skinner, 1991) and, beneath it, the synthesis of a new one (Stringfellow and Skinner, 1988). Although emergence of the animal at ecdysis is accomplished in a few minutes or hours, preparations for ecdysis may take days or months, and it is during this preparatory period that regeneration of lost limbs or antennae takes place. The focus in this discussion is on regeneration of limbs.

An outstanding property of the higher Crustacea is autotomy, or self-amputation. When a crustacean limb is injured or even firmly grasped in a life-threatening situation, the animal can cast off one or more limbs at a preformed breakage plane in the basiischial joint near the base of each limb (Figure 3.2; Hodge, 1958). This is accomplished by the contraction of specialized muscles that arise and insert on either side of the autotomy plane. The muscles do not cross the joint and therefore are not injured during autotomy, as are the major nerves and artery that serve the limb. The latter structures pass through an opening in a diaphragmlike autotomy membrane, and autotomy is accompanied by the closing of this opening, leaving the membrane sealed over the self-inflicted breakage. Although Crustacea have a clotting mechanism in their hemolymph (blood), they also have an

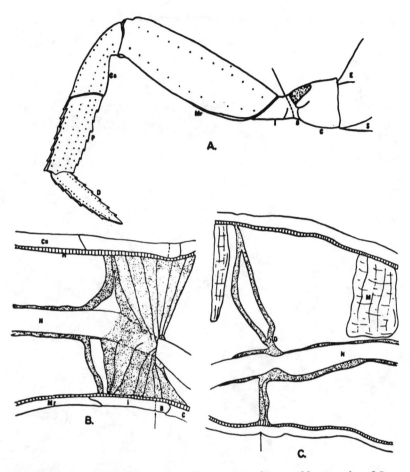

Figure 3.2. Intact walking leg (pereopod), autotomy plane, and leg muscles of *Gecarcinus lateralis*. (*A*) (*l*) Breakage plane at which limb separates from its base during autotomy. (*B*) Internal ventral view of autotomy plane (*arrow*). Stippled area is connective tissue sheath and autotomy diaphragm. (*C*) Lateral view of autotomy region. *Arrow* indicates autotomy plane. Autotomy diaphragm, *D*, traversed by muscle *M*. (Hodge, 1958)

open circulation. Breaks in the exoskeleton carry the threat of extensive loss of hemolymph, which the autotomy membrane prevents.

Réaumur (1718) described with great accuracy the process of ecdysis (Figure 3.3), as well as regeneration. During proecdysis, the regenerating limb was not externally visible, except as a sac attached to the animal, but he dissected the sac and found the growing limb within (Figure 3.4). This observation influenced Lazzaro Spallanzani's subsequent interpretation of regenerating blastemas of salamander limbs.

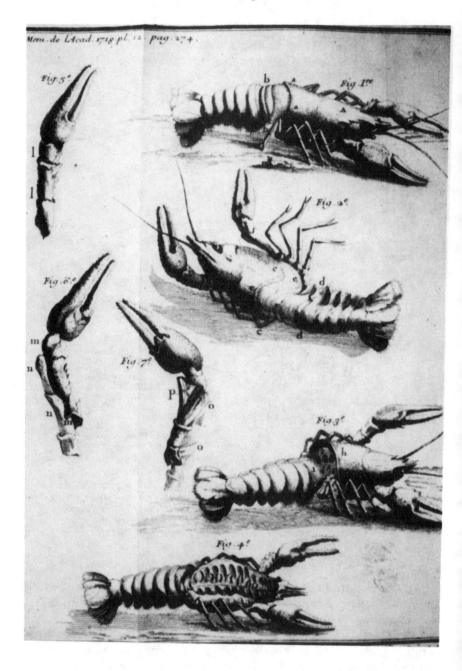

Réaumur also noted that frequently his attempts to excise a portion of a limb led to autotomy (although he did not call it that), and he described the position of the autotomy plane and the appearance of the autotomy membrane. He noted that a limb amputated at the autotomy plane regenerated more rapidly than limbs cut elsewhere, even though the amount that had to be regenerated was greater from the autotomy plane. And he contrasted regeneration with the development of a fetus in that regeneration does not occur within a fixed time frame but, "like plants," occurs according to the season and is more rapid in warmer weather. Although he did not make the connection, in most adult Crustacea molting and hence regeneration occur only in warm weather.

In his 1712 paper (p. 227) he states, significantly, "Nature gives back to the animal precisely and only that which it has lost, and she gives back to it all that it has lost." This sentence is a concise summary of his experiments. He had observed that although in the wild a wounded limb was cast off at the basal joint closest to the body, experimentally a cut could be made that excised no more than the tip of the claw. In such cases, the regenerate perfectly replaced the missing part. According to the preformation theory, this result meant that there had to be an "infinity of eggs" that filled the limb at every point, since otherwise the regenerated structure would be the entirety of the structure contained in the next "egg" close to the point of excision. Such aberrant regenerates were not observed.

At another time Réaumur had studied polydactyly (possession of extra toes and fingers) in humans, and he was aware that children had physical characteristics of both parents. After all these experiences, he was not enthusiastic about preformation theories.

Among the delights in Réaumur's regeneration paper are his whimsical comments and philosophical speculations. He notes that if humans could autotomize, that would be a boon to purse snatchers who were seized in the act. And if humans could regenerate limbs, he

Figure 3.3 Réaumur's illustrations of ecdysis in crayfish (numbered *1–7* in original plate). (*1*) Animal has detached first segment of exoskeleton and the dorsal carapace. (*2*) Animal is ready to exuviate, edge of carapace is elevated. (*3*) Exuvia after ecdysis. (*4*) As in *3*, except carapace has been lifted up in order to expose segments that remain attached to exuvia. (*5*) Part of normal limb. The letters (*l, l*) indicate sites of joints between second and third articulations. (*6*) Same limb at stage when muscle mass that occupied the largest joint (propus) has reached narrow point marked *l, l* in *5*. Exoskeleton is ruptured at suture (*mm*) exposing muscle mass. (*7*) Same limb shown in another position, providing better view of exoskeleton, which is open at *OO*, exposing same tissue as at *mm* in *6*. (Réaumur, 1718; courtesy of the Bancroft Library, University of California, Berkeley)

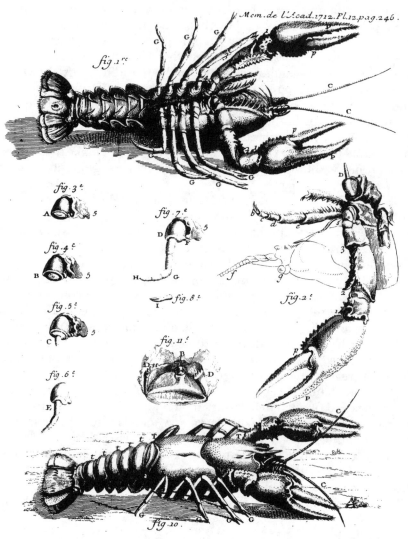

Figure 3.4. Réaumur's illustrations of regeneration in crayfish (numbered 1–11 in original plate; number 9 is missing in original). (*1*) Crayfish, lying on back to permit visualization of pereopods. (*2*) Anterior end of crayfish (enlarged), again lying on back, showing normal left claw and a regenerating right claw but only pale outline of anterior carapace. (*3–7*) Successive stages of regenerating limb bud. Numeral 5 on these figures indicates basal joint, considered fifth joint by Réaumur but now recognized as seventh. (*3*) *A* indicates end of limb, i.e., autotomy plane (see Fig. 3.1). (*4*) *B* indicates end of limb when autotomy membrane that covers it has become convex as regenerating papilla forms. (*5*) *C* denotes fleshy cone that regenerates to replace

himself might be more willing to bear arms in combat. In any case, he knew better. He writes,

Do we not have cause to complain to Nature who has treated more favorably than us animals that appear to us so vile? No. But Nature has provided us with a beautiful opportunity to admire her foresight. She has given to crayfish, and to all animals of that type, long limbs rather than hands; she has made them large at the extremities and slender at their origins. As it must be with such a structure, and the shell that covers it, that they break easily near their articulations, she has placed these animals in a state to repair their loss ... in the time when they can make only those movements demanded by the needs of their life; we have nothing similar to fear. (pp. 225–6)

This idea that regeneration is adapted to vulnerability appealed to Réaumur. He knew that tails of lizards regenerated, and, perhaps not surprisingly, he equated the abdomens of crayfish with "tails." Excised abdomens, he found, do not regenerate; crayfish without their hind parts die within a day or two. He used the vulnerability argument to explain that crayfish do not regenerate "tails" because their tails are large and powerful, and hence less vulnerable than lizard tails. Nevertheless, as Morgan (1898) showed in experiments on hermit crabs, liability to injury does not necessarily correlate with ability to regenerate.

Réaumur also remarked on the mechanical problem of withdrawing the largest segment of the claw (propus) through the narrow basal joint (basiischium) at ecdysis. Herrick (1895), in one of his several thorough treatises on the American lobster, graphically represented the problem (Figure 3.5), noting that the propus is several times larger than the basiischium. Some two hundred fifty years after Réaumur, we noted that one way Crustacea (at least Bermuda land crabs) solve this problem is to degrade muscle; there is more than a 40 percent reduction in the mass of muscle in the claw of the Bermuda land crab during proecdysis (Skinner, 1966, p. 118). Years later, work in our laboratory demonstrated that during proecdysial atrophy, there are areas of erosion in muscle fiber and a preferential loss of proteins that comprise the thin filaments (Figure 3.6; Mykles and Skinner, 1981, 1982a). The calcium-dependent proteinases responsible for muscle

Caption to Figure 3.4 *(cont.)*
missing leg. (*6*) Same cone as it begins to curve. (*7*) *FGH*, limb ready to emerge. It is curved at *G*, as is normal limb. (*8*) Separated pincers can be seen through transparent membrane. (*10*) Intermolt crayfish. (*11*) Stomach, containing three teeth (gastroliths) and cartilages that suspend them. (Réaumur, 1712. Courtesy of the Bancroft Library, University of California, Berkeley.)

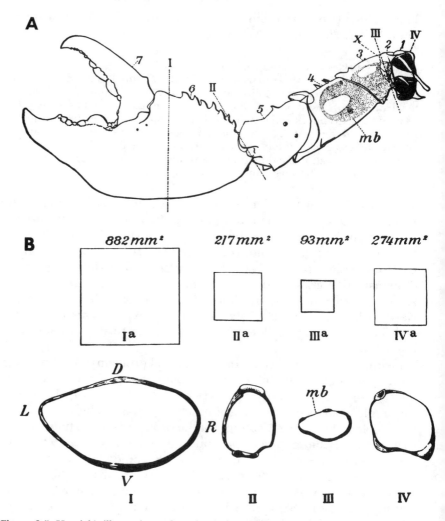

Figure 3.5. Herrick's illustrations of mechanical problems encountered during exuviation in lobster *Homarus americanus*. (*A*) Dorsal view of left chela (Arabic numbers 1–7 indicate segments of limb; Roman numerals I–IV indicate positions of cross sections shown in *B*. (*B*) Transverse sections of chela at positions I–IV. (*Top, Ia-IVa*) areas of sections expressed graphically and numerically. (*Bottom, I–IV*) cross sections. At *mb*, in *A* and *B*, calcium salts have been resorbed, permitting exoskeleton to distend, thereby enlarging cross-sectional area at position *III*. (Herrick, 1895)

Figure 3.6. Normal and atrophying claw muscles in *Gecarcinus lateralis*. Transverse sections of closer muscle fibers from *A* intermolt and *B* premolt animals. (*A*) Parts of three normal myofibrils and surrounding sarcoplasmic reticulum, sectioned through A-band. (*B*) Atrophying myofibrils with areas of erosion in myofilaments (*open arrowheads*). Note enlarged spaces between fibrils (∗). *D*, dyad; *Mf*, myofibril; *SR*, sarcoplasmic reticulum. *Bar*, 1 μm. (Mykles and Skinner, 1981, Fig 1.b, p. 317; Fig 2.a, p. 318)

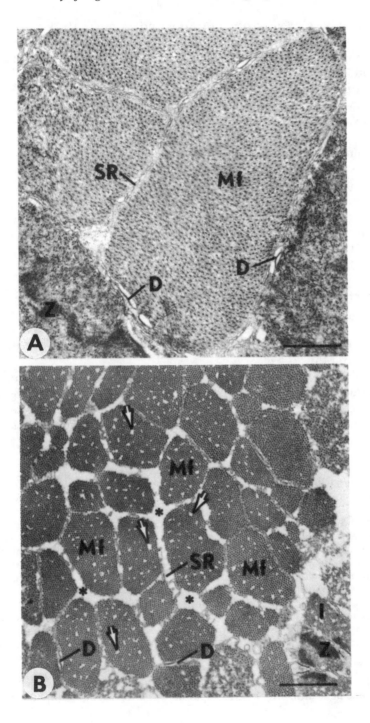

atrophy have been purified (Mykles and Skinner, 1983, pp. 10474–80). As these examples reveal, current research is still profitably investigating a biological problem clearly enunciated many years earlier by astute observers like Réaumur and Herrick.

It appears that the first systematic, rather than merely anecdotal, observations on regeneration apparently were those, such as Réaumur's, published in the eighteenth century. Regeneration aside, autotomy in crustaceans was well known in the eighteenth century and certainly must have been known for centuries before that. In 1777, Travis wrote, "Lobsters fear thunder, and are apt to cast [off] their claws on [hearing] a great clap. I am told that they will do the same on [hearing] the firing of a great gun, and that when men-of-war meet a lobster boat a jocular threat is used, that if the master does not sell them good lobsters they will salute him." In recounting this tale, Herrick (1895, p. 103) considered the autotomy response to a loud noise an unlikely fable. Earlier, however, Bell (1853) quoted J. E. Saunders, a "respectable fish salesman of Thames Street," as observing that lobsters "shoot [cast off] their claws, especially after a thunderstorm or the report of a cannon, and [the profits of] whole voyages are destroyed by this means. If time were given, new claws would be formed. It is a voluntary act, and does not injuriously affect the animal" (p. 245).

Bell (1853, p. 246) described remarks made by "his observant and accurate friend, Mr. R. Q. Couch," who "speaks of the effects of injuries to the antennae, and observes that it is an erroneous opinion that these organs are ordinarily thrown off in consequence of violence done to them, and afterwards renewed."

I have not [Couch proceeds] found this to be the fact; but, subjecting the parts to blows or fracture, both in short and long-tailed Crustaceans, I have found the creature suffering acutely from the injury, most so when just emerged from the water; but in no case have they rejected the whole organ in consequence of the violence. If, however, it be violently handled, a separation takes place at the terminal joint of the peduncles, in preference to any other place; and from this wound no stream of blood flows, but a fine membrane quickly forms on the surface, by which all effusion is prevented. This preservative process resembles that which takes place in case of the loss of the legs, and for the same purpose; for crabs and lobsters soon bleed to death, if the haemorrhage be not restrained. It is only the legs, including those bearing the chelae or nippers, that are readily and willingly thrown off by the animal; and in some cases, as in *Porcellana platycheles* [a hermit crab], this is not only done on the infliction of violence, but as if to occupy the attention of some dreaded object, while the timid creature escapes to a place of safety. The general method of defence is to seize the object with the pincers, and while these are left attached, inflicting, by their spasmodic twitchings, all the pain they are able to give, the crab, lightened of so great an incumbrance, has sought shelter in its hiding-place. It is by the short and quickened muscular action of the limb itself, and not by any effort of the body or peduncle, that this is effected; as the convulsion will continue for a considerable time after the separation, it follows that the twisting off of the claw, where the animal has seized human flesh for instance, or any other sensible object, is the direct way to increase the violence of the grasp. [Both of the present authors have, however, had the painful experience of having an autotomized claw attached to a human finger. Following autotomy, the former owner of the claw indeed scuttled to safety, leaving the pained investigator to disengage the abandoned claw, still quivering in contracture.] Any or all the legs may be thrown off on the receipt of injury, but not with equal facility in all the species; for in some, as in the common crab, if they be crushed or broken without great violence, they are sometimes retained, and the creature will in no long time bleed to death. To save the crab the fisherman proceed to twist off the limb at the proper joint, or give it a smart blow, when it is rejected; and in either case the bleeding is stopped. Fracture of the crust at the extreme points of the legs is not much regarded; for, being filled with an insensible cellular membrane, no violent action is excited in the muscular structure, and the part seems capable of some attempt at restoration, at least sufficient to render the evil endurable until the period of a general renewal of the surface.

After the loss of a limb, a considerable time elapses before any attempt at restoration is visible: but under some circumstances the process is much accelerated; and while it is advancing, it is commonly found that the flesh of the creature is *unusually flaccid and watery*. [Emphasis added. This may be among the earliest observations of muscle atrophy during proecdysis (the premolt period).] In the most common species, the first appearance of the new limb is in the middle of the scar, doubled on itself, but with all the proper proportions, and enclosed in an exceedingly fine membrane, by which it is bound down. Much of the first stage of the growth of the new limb is accomplished before it acquires density; but when the crust is rendered firm, the nutrition no longer proceeds through the encasing membrane; which a slight motion of the limb lacerates, and the leg extends to its natural position; but it continues for a long time of a much smaller size than the corresponding one of full growth, sometimes also appearing as if distorted, either from deficient nourishment, or from injury received in its unprotected state.

Bell continues with his own observations:

I have omitted from this interesting detail some speculations of the observant author, and some statements respecting which he himself speaks doubtfully; and it appears to me that it contains by far the most satisfactory and most simple statements of this interesting fact that have ever appeared. Although Mr. Couch's observations were chiefly made upon brachyurous [true crab] forms, there is no doubt that the process is precisely similar in all the higher forms of Crustacea.

As microscopes improved throughout the nineteenth century, so did the histological observations on molting and regeneration, and the number of detailed phenomenological descriptions increased. That regenerates arise from germ cells positioned at the base of the limb was maintained midway in the century, when H. D. S. Goodsir (1844) described to the Wernerian Natural History Society a "small glandular-like body" in the basipodite of each crustacean limb, which "supplies the germs of future legs [and] contains a great number of large nucleated cells ... interspersed throughout a fibro-gelatinous mass. A single branch of each of the great vessels, accompanied by a branch of nerve, runs through a small foramen near the center of this body, but there is no vestige either of muscle or tendon, the attachments of which are at each extremity." Herrick (p. 107) was not able to substantiate these observations and called them simply erroneous. Of major importance in Herrick's work was that he was looking in various sections of limbs specifically for such embryonic cells that could be the source of a new limb but found only fully differentiated, normal tissue cells. "Embryonic cells may be present but are not discernible," he wrote (p. 108). Preformation dying hard, he also looked for and failed to find "any preformed organ" (p. 108). He suggested that the limb formed from "growth of connective tissue cells already present in the stump" (p. 107). That the regenerate grows out from preexisting cells was unquestioned and was not the issue; the question was whether there exist homunculuslike limbs, and parts of limbs, capable of expanding into a full-size replacement of the missing part. After the acceptance of the cell theory and Virchow's famous *Omnis cellula e cellula* ("Every cell comes from a cell."), the preformation theory was less compelling, although the process of growth and regeneration was not much less mysterious.

In 1892 Léon Frédéricq investigated autotomy in the green crab *Carcinus maenas* and found it to be a reflex mediated in a bundle of nerve cells, the thoracic ganglion; the response remained intact after removal of the supraesophageal ganglion, which was considered to be the analog of the brain and the seat of volition. This led to the conclusion that autotomy is an involuntary response to an adequate stimulus, rather than a willful decision on the part of the crab. We leave to others the issue of whether green crabs have Free Will over any aspect of their behavior.

A remarkable regeneration phenomena, still under investigation, was described by Przibram in 1901. This is asymmetry reversal, as seen in the snapping shrimp *Alpheus*. This animal has a large ("snapper") claw and a smaller ("cutter") claw (Figure 3.7). If the cutter claw is lost, it regenerates at the next molt. But if the snapper is lost, at the next molt the cutter becomes the snapper and the snapper the cutter. If both are lost, the switch does not occur. Wilson (1903) showed that the

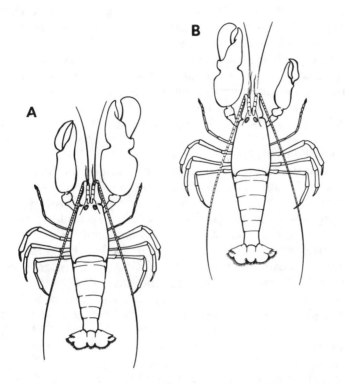

Figure 3.7. Reversal of symmetry. (*A*) *Alpheus,* snapping shrimp, has small cutter claw (*left*) and large snapping claw (*right*). If cutter claw is lost, another cutter regenerates at next molt. (*B*) If snapping claw is lost, at next molt cutter becomes snapping claw, and lost snapping claw becomes cutter. (Goss, 1969)

reversal could be prevented if, when the snapper is removed, the nerve supply to the cutter is severed. Histological examination revealed no obvious differences in the nerve supplies of the claws or the ganglia from which they originate. Many crustacean species have asymmetric claws, and it has been found that reversal is not a general phenomenon. For example, male fiddler crabs have one claw much larger than the other. Charles Zeleny (1905a) found that if the large claw is removed from a male fiddler crab *Gelasimus (Uca) pugilator,* a large claw regenerates on the same side of the animal as before.

Zeleny, in 1905, experimented on invertebrates from several phyla, as well on plants, to discern the effect of the presence or absence of one part of an organism on the development of another. Zeleny believed that much could be learned from observing regeneration in "mutilated" (1905b, p. 347) organisms (that is, those from which one or more parts have been removed). In the course of his investigations, he removed the eyestalks of fiddler crabs and noted that the few animals that survived molted more rapidly than expected. Removal of both eyestalks resulted in molting sooner than removal of a single eyestalk. The eyes themselves did not regenerate, and in fact three of the five animals died soon after the operation. It was to be nearly another forty years before the significance of this eyestalk effect was appreciated.

Zeleny (1905b) also appears to have been the first to record data showing that the loss of limbs leads to accelerated molting; autotomy of two claws is more effective than autotomy of only one. We and others have now observed this response in fifteen crustacean species. The response is useful for experiments. If, for example, an anecdysial (intermolt) *Gecarcinus lateralis* (Bermuda land crab) is induced to autotomize five or more walking legs, it is propelled almost immediately into proecdysis, during which all the missing legs are regenerated. By this means the initiation of molt preparations and of regeneration can be controlled by the investigator in the laboratory. An added bonus is that the ensuing ecdysis, in contrast to that following eyestalk removal, is normal, and virtually 100 percent of the crabs survive. The stimulus to this precocious molt is unknown. We tentatively refer to it as LAF_{an}, an acronym for "limb autotomy factor, anecdysis."

The effects of eyestalk removal on molting were rediscovered more or less simultaneously by Brown and Cunningham (1939), Ralph Smith (1940), and the Abramowitzes (1940). None of them seem to have known Zeleny's work. They found that, as in "normal" molting periods, regeneration of limbs is stimulated by removal of the eyestalk. In any event, the observation this time came at a more propitious moment, because, as a result of the work of Perkins (1928), Koller (1930), Kleinholz (1938), and others, the eyestalk was by then recognized as the source of chromatophorotropic hormones, and the X-organs and sinus glands in the eyestalk had been described by Hanström (1931, 1933). (The name "X-organs," incidentally, came from a note on one of Hanström's drawings, where he marked a group of interesting cells with an X.) Since eyestalk removal accelerated molting, it seemed likely that the sinus gland in the eyestalk secretes a molt-influencing substance, specifically a molt-inhibiting hormone. This was supported by Brown and Cunningham's replacement experiments in which they showed that implantation of sinus glands into operated animals inhibited molting. A good deal of research is currently being done on the isolation and purification of

MIH (Mattson and Spaziani, 1985; Webster, 1986; Chang, Bruce, and Newcomb, 1987), but a detailed discussion of that work belongs in a review of current studies rather than in this discussion of historical research.

The principal, and possibly the only, effect of MIH is to inhibit the secretion of ecdysteroids by the Y-organs. These paired organs were recognized as a common feature of the malacostracans by Gabe (1953, 1956) and were named "Y-organs" to pair with the already-named X-organs. By analogy to the prothoracic glands of insects, Gabe suggested (correctly, it turned out) that secretions of the Y-organs stimulate molting; and, of course, their secreted ecdysteroids also stimulate regeneration.

The interactions between regeneration and molting are complex. Following Zeleny's (1905b) lead, and expanding on R. Q. Couch's description (quoted in Bell, 1853) of the regenerating limb folded upon itself, we have observed the following sequence in *Gecarcinus*. After the induction of autotomy of five or more walking legs, an animal promptly enters proecdysis (premolt), a period lasting about eight weeks (Skinner and Graham, 1970). During this time the regenerating limbs can be clearly seen and their growth measured. An extraordinary event occurs if one of these partial regenerates is autotomized before a specific stage in proecdysis (ibid., p. 232). The level of circulating ecdysteroids plummets (Hoarau and Hirn, 1981, p. 96), and molting preparations in the remainder of the animal are put on hold while a secondary regenerate grows at the autotomy plane (Figure 3.8). This makes sense if one remembers that the earliest stage of regeneration, formation of a papilla at the autotomy plane, occurs during anecdysis at a time when the level of ecdysteroid is low. After the secondary regenerate catches up with its fellows, molt preparations resume again, and a normal ecdysis, delayed by the "catch-up" period, ensues (Holland and Skinner, 1976). The point of interest here is that, in part of the animal, molting and regeneration are inhibited while, in another part, formation of a second regenerate is stimulated. We have postulated that the interruption is caused by an unidentified factor, which we refer to as LAF_{pro}, for "limb autotomy factor, proecdysis." There is a critical point in proecdysis, about three weeks before the anticipated ecdysis, after which this compensatory process does not occur and preparations for ecdysis are not delayed by amputating a partially regenerated leg. (We now think that the critical point is the onset of stage D_1, when the epidermis separates from the membranous layer of the old exoskeleton, which is being enzymatically degraded [Skinner and Graham, 1972; O'Brien and Skinner,

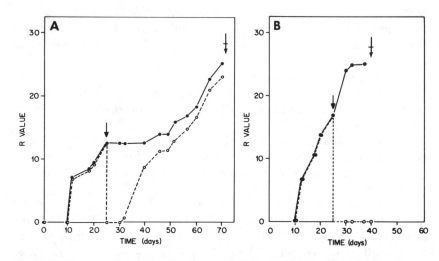

Figure 3.8. Patterns of growth of regenerating limbs in *Gecarcinus lateralis*. Day 0: animals caused to autotomize 8 pereopods; regenerating limbs were measured and *R* values calculated (*R* value = length of regenerate [cm.] ÷ carapace width [cm.] × 100; Bliss, 1956, p. 58). (*A*) At (↓), intermediate-size regenerate (*R* = 12, *open circle*) was removed. Growth of primary regenerate (*closed circle*) that remained on animal stopped for ~ 18 days. Papilla of secondary regenerate (*open circle*) that would grow to replace lost primary regenerate appeared on ~ day 32 and grew rapidly. When size of secondary regenerate was approximately equal to that of primary regenerate, growth resumed in latter. Overall, time to ecdysis (↓) was prolonged from ~ 55 days to 70 days. (Control not shown.) (*B*) At (↓), large regenerate (*R* = 17) was autotomized. No secondary regenerate appeared, growth of primary regenerate was not inhibited, and ecdysis was not delayed; it occurred on day 40 (↓). (Holland and Skinner, 1976)

1987, 1988].) After the critical point, only a papilla forms at the site, and the limb does not regenerate until the following proecdysial period.

But this description again strays into current research, not history. Our purpose in this essay has been to present a crab's-eye view of how the early work has led to present research on regeneration and its interrelations with molting. The subject remains as mysterious and fascinating as ever.

ACKNOWLEDGMENTS

We thank Dr. B. Bang and Ms. J. Fessenden, of the Marine Biological Laboratory Library, and Dr. M. Rossiter for locating source material, and Dr. D. L. Mykels and Dr. R. I. Smith for providing Figures 3.3. and 3.4 from the

Bancroft Library of the University of California, Berkeley. We thank Dr. M. H. LaPrade for permission to include Figure 3.2 from her Ph.D. thesis (M. H. Hodge, "Some Aspects of the Regeneration of Walking Legs in the Land Crab *Gecarcinus lateralis,*" Harvard University, 1958). Research by Dorothy Skinner and her trainees was supported by grants from the National Science Foundation, a Training Grant from the National Institutes of Health, a postdoctoral fellowship from the Muscular Dystrophy Association, and the Office of Health and Environmental Research, U.S. Department of Energy under contract DE-AC05-84OR21400 with Martin Marietta Energy Systems, Inc.

REFERENCES

Abramowitz, R. K., and A. A. Abramowitz. 1940. Moulting, growth and survival after eyestalk removal in *Uca pugilator. Biol. Bull.* 78: 179–88.

Bell, T. 1853. *A History of the British Stalk-eyed Crustacea,* pp. 246–8. Van Voorst, London.

Bliss, D. E. 1956. Neurosecretion and the control of growth in a decapod crustacean. In *Bertil Hanstrom: Zoological Papers in Honour of his Sixty-fifth Birthday* (K. G. Wingstrand, ed.), pp. 56–75. Zoological Institute of Lund, Lund, Sweden.

Bliss, D. E., and J. H. Welsh. 1952. The neurosecretory system of brachyuran Crustacea. *Biol. Bull.* 103: 157–69.

Brown, F. A., and O. Cunningham. 1939. Influence of the sinus gland of crustaceans on normal viability and ecdysis. *Biol. Bull.* 77: 104–14.

Chang, E. S.; M. J. Bruce; and R. W. Newcomb. 1987. Purification and amino acid composition of a peptide with molt-inhibiting activity from the lobster, *Homarus americanus. Gen. Comp. Endocrinol.* 65: 56–64.

Frédéricq, L. 1892. Nouvelles recherches sur l'autotomie chez le crabe. *Arch. Biol.* 12: 169–97.

Gabe, M. 1953. Sur l'éxistence, chez quelques Crustacés Malacostracés, d'un organe comparable à la glande de la mue des insects. *C. R. Acad. Sci. Paris* 237: 1111–13.

  1956. Histologie comparee de la glande de mue (organe Y) des Crustacés Malacostracés. *Ann. Sci. Nat. Zool. et Biol. Animale* 18: 145–52.

Goodsir, H. D. S. 1844. A short account of the mode of reproduction of lost parts in the Crustacea. *Ann. Mag. Nat. His.* 13: 67.

Goss, R. J. 1969. *Principles of Regeneration.* Academic Press, New York.

Gough, J. B. 1975. "Réaumur, René-Antoine Ferchault de," *Dictionary of Scientific Biography,* 11: 327–35.

Hanström, B. 1931. Neue Untersuchungen über Sinnesorgane und Nervensystem der Crustacean. I. *Zeitschr. f. Morphol. u. Okol. d. Tiere* 23: 80–236.

  1933. Neue Untersuchungen über Sinnesorgane und Nervensystem der Crustacean. II. *Zool. Jahrb., Abt. Anat. Ont. d. Tiere* 56: 387–520.

Herrick, F. H. 1895. *The American lobster: A study of its habits and development. Bull. U.S. Fish Comm.* 15: 1–252.

  1909. Natural history of the American lobster. *Bull. Bur. Fish.* 29: 149–408.

Hoarau, F., and M. Hirn. 1981. Effects of amputation and subsequent regeneration of a leg on the duration of the intermolt period and the level of circulating ecdysteroids in *Helleria brevicornis* Ebner (ground isopod). *Gen. Comp. Endocrinol.* 43: 96–104.

Hodge, M. H. 1958. Some aspects of the regeneration of walking legs in the land crab *Gecarcinus lateralis.* Ph.D. thesis, Harvard University.

Holland, C. A., and Skinner, D. M. 1976. Interactions between molting and regeneration in the land crab. *Biol. Bull.* 150: 222–40.

Kleinholz, L. H. 1938. Studies in the pigmentary system of Crustacea. IV. The unitary versus the multiple hormone hypothesis of control. *Biol. Bull.* 75: 510–32.

Koller, G. 1930. Weitere Untersuchungen über Farbwechsel und Farbwechsel-hormone bei *Crangon vulgaris. Z. Vergleich. Physiol.* 12: 632–67.

Mattson, M. P., and E. Spaziani. 1985. Characterization of molt-inhibiting hormone (MIH) action on crustacean Y-organ segments and dispersed cells in culture and a bioassay for MIH activity. *J. Exp. Zool.* 236: 93–101.

Morgan, T. H. 1898. Regeneration and liability to injury. *Zool. Bull.* 1: 287–300.

Mykles, D. L., and D. M. Skinner. 1981. Preferential loss of thin filaments during molt-induced atrophy in crab claw muscle. *J. Ultrastruct. Res.* 75: 314–25.

1982a. Molt cycle-associated changes in calcium-dependent proteinase activity that degrades actin and myosin in crustacean muscle. *Dev. Biol.* 92: 386–97.

1982b. Crustacean muscles: Atrophy and regeneration during molting. In *Basic Biology of Muscles: A Comparative Approach* (B. M. Twarog, R. J. C. Levine, and M. M. Dewey, eds.), pp. 337–57. Raven, New York.

1983. Ca^{2+}-dependent proteolytic activity in crab claw muscle: Effects of inhibitors and specificity for myofibrillar proteins. *J. Biol. Chem.* 258: 10474–80.

Needham, A. E. 1945. Peripheral nerve and regeneration in Crustacea. *J. Exp. Biol.* 21: 144–6.

1946. Peripheral nerve and regeneration in Crustacea. II. *J. Exp. Biol.* 22: 107–9.

1949. Depletion and recuperation of the local factors during repeated regeneration. *J. Exp. Zool.* 112: 207–31.

1953. The central nervous system and regeneration in Crustacea. *J. Exp. Biol.* 30: 153–9.

O'Brien, J. J.; S. Kumari; and D. M. Skinner. 1990. Proteins of the individual layers of the crustacean exoskeleton. Unpublished observations.

O'Brien, J. J., and D. M. Skinner. 1987. Characterization of enzymes that degrade crab exoskeleton. I. Two alkaline cysteine proteinase activities. *J. Exp. Zool.* 243: 389–400.

1988. Characterization of enzymes that degrade crab exoskeleton. II. Two acid proteinase activities. *J. Exp. Zool.* 246: 124–31.

Perkins, E. B. 1928. Color changes in crustaceans, especially in *Palaemonetes. J. Exp. Zool.* 50: 71–103.

Przibram, H. 1901. Experimentelle Studien über Regeneration. *Wilhelm Roux Arch. Entw. Mech.* 11: 321–45.

Réaumur, R. A. F. 1712. Sur les diverses reproductions qui se font dans les Ecrevisse, les Omars, les Crabes, etc. et entr'autres sur celles de leurs Jambes et de leurs Ecailles. *Mem. Acad. Roy. Sci.*, pp. 223–45.

1718. Aux Observations sur la Mue des Ecrivisses, données dans les Memoires de 1712. *Hist. Acad. Roy. Sci.,* pp. 263–74.

Skinner, D. M. 1966. Breakdown and reformation of somatic muscle during the molt cycle of the land crab, *Gecarcinus lateralis. J. Exp. Zool.* 163: 115–24.

1985a. Molting and regeneration. In *The Biology of Crustacea* (D. E. Bliss and L. H. Mantel, eds.), 9: 43–146. Academic Press, New York.

1985b. Interacting factors in the control of the crustacean molt cycle: Advances in crustacean endocrinology. *Am. Zool.* 25: 275–84.

Skinner, D. M., and D. E. Graham. 1970. Molting in land crabs: Stimulation by leg removal. *Science* 169: 383–5.

1972. Loss of limbs as a stimulus to ecdysis in Brachyura (true crabs). *Biol. Bull.* 143: 222–33.

Smith, R. I. 1940. Studies on the effects of eyestalk removal upon young crayfish (*Cambarus clarkii* Girard). *Biol. Bull.* 79: 145–52.

Stebbing, T. R. R. 1893. *A History of British Crustacea.* Kegan, Paul, Trench, & Trubner, London.

Stringfellow, L. A., and D. M. Skinner. 1988. Molt-cycle correlated patterns of synthesis of integumentary proteins in the land crab *Gecarcinus lateralis. Dev. Biol.* 128: 97–110.

Travis, D. F. 1960. The deposition of skeletal structures in the Crustacea. I. The histology of the gastrolith skeletal tissue complex and the gastrolith in the crayfish, *Orconectes (Cambarus) virilis* Hagen – Decapoda. *Biol. Bull.* 118: 137–49.

Webster, S. G. 1986. Repression of Y-organ secretory activity by molt-inhibiting hormone in the crab *Pachygrapsus crassipes. Gen. Comp. Endocrinol.* 61: 237–47.

White, A. 1857. *A Popular History of British Crustacea.* Lovell Reeve, London.

Wilson, E. B. 1903. Notes on the reversal of asymmetry in the regeneration of the chelae in *Alpheus heterochelis. Biol. Bull.* 4: 197–210.

Zeleny, C. 1905a. Compensatory regulation. *J. Exp. Zool.* 2: 1–102.

1905b. The relation of the degree of injury to the rate of regeneration. *J. Exp. Zool.* 2: 347–69.

4

Abraham Trembley and the origins of research on regeneration in animals

HOWARD M. LENHOFF and SYLVIA G. LENHOFF

The first operation I performed on the polyps [November 25, 1740] was to cut them transversely. . . . I place it with a little water in the hollow of my left hand. . . . When I have it as I want it, I delicately pass one blade of the scissors, which I hold in my right hand, under the part of the polyp's body where it is to be severed. Then I close the scissors. (Abraham Trembley, 1744b)

Thus, with a single snip of the scissors across the middle of an extended hydra in November 1740, the young Swiss tutor Abraham Trembley initiated the modern study of regeneration.

Since the time of Aristotle it had been known that certain creatures can replace a lost limb. Fishermen and folk living beside the sea had long observed that in some cases a severed limb, such as the arm of a starfish, could regenerate to form a complete animal. As early as 1714, the great French naturalist René-Antoine Ferchault de Réaumur published descriptions of his conclusive experiments showing that crayfish can replace a missing appendage. (See Chapter 3 of the present volume. For a more detailed discussion of early reports of regeneration, see Chapter 6.)

Why, then, did scholars pay scant attention to the earlier anecdotal reports of regeneration and to Réaumur's studies. Regarding the earlier reports, Trembley stated that academicians did not themselves attempt to observe the phenomenon, and in regard to their lack of interest in his own dramatic discoveries, he judged that they did not want to believe what did not agree with their own views of nature. In Trembley's words, because of accepted views of generation they "presumed [regeneration of a complete animal from a piece of an animal] impossible" (Trembley, 1986, p. 186). Trembley himself vehemently opposed general rules based on theory rather than on observation. He chastised the cloistered theoretician: "Notice that it was only uneducated people who could not have been imbued with the prejudices of the schools, who quite simply believed based on the facts they

47

observed that parts of an animal can become complete animals" (Trembley, 1986, p. 187).

In this chapter, we (1) briefly sketch Trembley's life and some of his scientific discoveries; (2) describe (mostly in his own words) how he made his discovery of regeneration; (3) provide a detailed summary of all his experiments on regeneration; (4) offer a chronology of how his discoveries on regeneration were first made known to the scientific world, doubted, confirmed, and eventually lauded; and (5) briefly discuss the influence of Trembley's discovery of regeneration on the thinking of Enlightenment scientists, philosophers, and theologians.

Who was Abraham Trembley?

Abraham Trembley (Figure 4.1) was an eighth-generation member of a prominent Geneva family (Baker, 1952). Born in 1710, he grew up in an age when many intellectuals in Geneva were turning their attention to natural history (Dawson, 1987). He himself began as a student of mathematics. At the Calvin Institute, now the University of Geneva, he prepared a thesis on the calculus (Trembley, 1730).

When he had finished his education, the young Trembley sought employment in Holland. He eventually became the tutor of the two young sons of Count William Bentinck of The Hague (Baker, 1952). While teaching the boys natural history, Trembley rediscovered the freshwater hydra, first described – independently and briefly – in 1704, by both Leeuwenhoek and an "anonymous Gentleman" from England (1704). With this animal, which Trembley called a "polyp of fresh water with arms shaped like horns" (1986), he began a series of investigations that rocked the scientific world, initiated the systematic study of experimental morphology, and set a standard for experimental biological research that could serve us well today.

The results of his experiments on hydra were published in full in his *Mémoires* in 1744(b). In addition to describing his experiments showing that complete animals can regenerate from small cut pieces of a hydra, Trembley also reported in those *Mémoires* for the first time that (1) animals can reproduce asexually by budding; (2) tissue sections from two different individuals of the same animal species can be grafted to each other; (3) the materials oozing out of the edges of cut tissue have certain properties (these fit the definition of protoplasm as described by Dujardin a century later); (4) living tissues can be stained ("vital staining") and those stained tissues can be used in experiments; and (5) eyeless individuals can exhibit a behavioral response to light ("positive phototaxis"). In addition, Trembley made more than fifty

Figure 4.1. Portrait of Abraham Trembley. The original etching was published as the frontispiece to Trembley's *Instructions d'un père à ses enfans, sur le principe de la vertu et du bonheur* (Chirol, Geneva, 1783). (It is signed "Clemens del: & Sculps, 1778.")

observations and discoveries on hydra that still hold (Lenhoff and Lenhoff, 1986).

Those discoveries were made in the short span of about four years, from 1740 to 1744. Moreover, Trembley carried out his experiments long before the development of sophisticated tools, relying mostly on a magnifying glass and occasionally on a simple microscope (Trembley, 1986).

How Trembley discovered regeneration

It is often stated that Trembley first cut a hydra into two pieces in order to determine whether his "polyp" was a plant or an animal. Although that is not exactly accurate, it is understandable that some historians have supposed so. The first hydra Trembley discovered was the green hydra. Because it did not resemble any known animal, and because it seemed to have some properties resembling those of plants, Trembley was not certain at first whether it was a plant or an animal. But soon after beginning his observations he became convinced that hydra were animals, as he saw them contract and extend and eventually "take steps much in the same way as do inchworms." Before beginning his detailed studies on hydra, Trembley states, "The sight of this step-by-step movement of the polyps finally persuaded me that they were animals. Once convinced, I stopped observing them for I had found what I was seeking" (1986, p. 6). In fact, it was this behavioral characteristic of taking steps that Trembley later used to determine whether or not a regenerated piece of a hydra had regained the properties inherent in an intact animal.

Why, then, did Trembley decide to cut a hydra in half? He did so because he had noticed a most unusual developmental feature: Not all the hydra had the same number of arms. Because such irregularity in number of limbs is unusual among animals, Trembley felt he would be remiss in not trying an experiment to prove once and for all that his polyps were animals. The experiment he chose was to cut a hydra in half and to wait to see if cut pieces would regenerate as some plants do. Because he was convinced that his polyps were animals, and because animals were not known to regenerate complete individuals, he felt certain that the cut pieces would die. But they did not, and he then commenced the first systematic study of regeneration.

To convey the genius of Trembley and the excitement of that initial discovery, we present here in translation Trembley's own words:

It was not long before I noticed that not all the individuals of the species of polyps that I was observing have an equal number of arms or legs. I had reason to believe that there was nothing unnatural about these variations. Although I found no difficulty in accepting this difference among the individuals of a single species of animals, I nevertheless compared these arms at first with the branches and roots of plants, the number of which varies greatly among the individuals of the same species. At this point, I speculated anew that perhaps these organisms were plants, and fortunately I did not reject this idea. I say fortunately because, although it was the less natural idea, it made me think of cutting up the polyps. I conjectured that if a polyp were cut in two and if each of the severed parts lived and became a complete polyp, it would be clear that these organisms were plants. Since I was much more inclined to think of them as animals, however, I did not set much store by this experiment; I expected to see these cleaved polyps die.

On November 25, 1740 I sectioned a polyp for the first time. I put the two parts [Figs. 2 and 3 in Figure 4.2] into a shallow glass which contained water to the height of only about nine to eleven millimeters. In this way I could observe the parts of the polyp easily with a rather strong magnifying glass.

I shall describe elsewhere the precautions I took in performing my experiments on these cut polyps as well as the manner in which I set about cutting them. For now, suffice it to say that I cut the polyp transversely a little closer to the anterior than to the posterior end. Thus the first part was a little shorter than the second.

The instant I cut the polyp the two parts contracted so that at first they looked like no more than two small granules of green matter at the bottom of the glass into which I put them. As I have said, the first polyps I cut were green in color. The two parts extended the same day that I separated them. They were quite easy to distinguish from one another because the first had its anterior end bedecked with those fine threads which serve as the polyp's arms and legs, whereas the second had none at all.

The act of extending itself was not the only sign of life that the first part gave on the day it was separated from the other, for I saw it move its arms. On the following day, the first time that I came to observe it, I found it had changed its place, and shortly afterward, I saw it take a step. The second part remained extended as on the preceding day and in the same spot. I shook the glass a little to see if it was still living. When this movement caused it to contract, I concluded that it was alive. A short time later it extended anew. On the days that followed I saw the same thing occur.

Still I considered the movement of the two parts of a single polyp merely as signs of a feeble remnant of life, especially as regards the second part. Since I was presuming that the polyp was an animal, I expected its head to be located on the anterior part, as indeed it is. It seemed natural enough to me that the half composed of the head and a portion of the body could still live. I thought that the operation which I had performed had only mutilated the head part without essentially disrupting its animal economy. I compared this first part to a lizard which has lost its tail and which does not die from losing it. Indeed, again supposing that the polyp was an animal, I assumed that the second part was only a kind of tail without the organs vital to the life of an animal [Fig. 4]. I did not think that it could survive for long separated from the rest of the body. Who would have imagined that it would grow back a head! I was observing this second half to find out how long it would retain the remnants of life; I had not the least expectation of being a spectator to this marvelous kind of reproduction.

I observed these parts through a magnifying glass several times each day. On the morning of December fourth, the ninth day after cutting the polyp, I

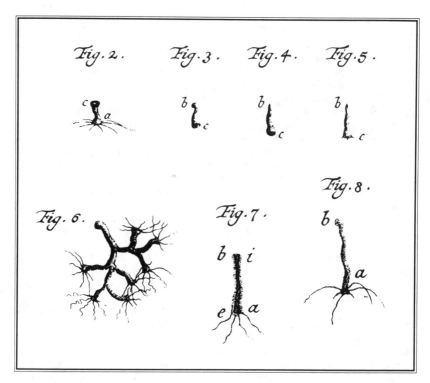

Figure 4.2. Illustrations of hydra regeneration from Trembley's *Mémoires* (1744). Although the text refers to the "polyp of the first species," the green hydra, all of these illustrations of regeneration, the only ones of this kind in Trembley's *Mémoires*, depict *H. vulgaris*, the "polyps of the second species." These illustrations were drawn and engraved by Pierre Lyonet.

Figs. 2 and 3: The "halves *ac* and *cb* of a polyp of the second species which has been cut in two. [Fig. 2] shows the first part: *a* is the head, *c* the place at which it has been cut." "[Fig. 3] . . . depicts the second part with its anterior end open" (Trembley, 1986, p. 188. The figures are figures 1 and 2 of Plate 11, *Memoir* 4.)

Fig. 4: The "second part *cb* is shown . . . , but its anterior end *c* is closed and a little swollen."

Fig. 5: The "second part *cb*. The arms are beginning to sprout at the anterior end *c*."

Fig. 6: "A Hydra with seven heads."

Fig. 7: "The half *aeib* of a polyp which has been cut lengthwise. This part has not yet closed. Its interior surface is shown. This half has four arms at its anterior extremity *ae*."

Fig. 8: "The same half as is shown in [Fig. 7], but with its edges entirely joined together. At this stage, this half already can be considered a complete polyp. The two new [short] arms can be seen beginning to grow in the spaces between the original set." (Items have been renumbered for ease of reference. Translations of the text of *Mémoires* are from Lenhoff and Lenhoff, 1986.)

thought I saw three small protuberances emerging from the edges of the anterior end of the second part, the one which had neither head nor arms [Fig. 5]. The moment I saw them I thought of the horns which serve the polyps as arms and legs. These protuberances were precisely where the arms would have been had this second part been a complete polyp. I did not want to conclude so quickly, however, that these were indeed arms that were beginning to grow. I continued to see these protuberances throughout the day, and I became extremely excited and impatient for the moment when I would know clearly what they were. Finally, on the next day they were large enough to dispel all doubt; these were truly arms growing at the anterior end of the second part. On the following day two new arms began to emerge, and a few days later three more came out. This second part then had eight arms which, in a short time, were as long as the arms of the first part, that is, those arms the polyp had before it was cut up.

From that time on I found no difference between the polyp that developed from the second part and a polyp that had never been cut up. The first part had given me that impression since the day following the operation. When I examined the two parts through a magnifying glass with all the attentiveness of which I was capable, both of them appeared to be demonstrably complete polyps performing all the functions of these organisms known to me: they elongated, contracted, and took steps. (Trembley, 1986, pp. 7–9)

The cut pieces had regained the very properties that had convinced Trembley earlier that the polyps were animals. Characteristically, Trembley takes no great credit for any genius behind his revolutionary experiment and discovery, but instead elegantly points out the role of serendipity in biological research:

Because of its nature, that finding was not to be the fruit of long patience and great wisdom, but a gift of chance. It is to such a happy chance that I owe this discovery which I made, not only without forethought, but without my ever having in my entire life any idea slightly related to it. (Ibid., p. 4)

Trembley's experiments

The experiments on regeneration of hydra that Trembley described in his *Mémoires* are summarized in Table 1 (reprinted here from Lenhoff and Lenhoff, 1986). A quick perusal shows that Trembley thought of virtually every possible way of sectioning hydra to investigate regeneration. He sectioned them transversely, longitudinally, transversely and longitudinally, and into many pieces; he sectioned through animals developing buds and then sectioned the parts, eventually creating a seven-headed "monster" (Fig. 6 in Figure 4.2), which he named "Hydra" after the monster of Greek mythology.

This seven-headed monster came into being after Trembley had cut a single polyp partially down the middle, starting at the head end,

Table 1. Abraham Trembley's Experiments On Regeneration Of Hydra

EXPERIMENT	RESULTS	ADDITIONAL OBSERVATIONS
I. TRANSVERSE SECTIONS:		
A. Hydra sectioned transversely across middle of body (i.e. into halves).	1. Piece with head regenerated new foot. 2. Piece with foot regenerated new head.	1. Rate of regeneration faster in warm weather. 2. Degree of nutrition may affect rate of regeneration positively.
B. Hydra sectioned transversely into three pieces (i.e. into thirds).	1. Piece with head regenerated new foot. 2. Middle piece regenerated new head and new foot. 3. Piece with foot regenerated new head.	
C. Hydra sectioned transversely into four pieces (i.e. into quarters).	1. Piece with head regenerated new foot. 2. Each of the two middle pieces regenerated a new head and a new foot. 3. Piece with foot regenerated a new head.	Some of the hydra that regenerated were fed and maintained for over two years along with animals that had never been sectioned. No differences could be noted between the two groups of hydra.
D. One hydra sectioned transversely into four pieces, the four hydra that regenerated from the four pieces were fed until each grew to full size, each of those four hydra were sectioned into three or four pieces, and so on until eventually fifty hydra were obtained by regeneration from the original one.	Eventually Trembley obtained fifty hydra from the one that was originally sectioned into four pieces.	It should be noted that the fifty hydra did not come from one hydra that was cut into fifty pieces initially, as is often erroneously stated.
E. A number of hydra sectioned transversely once, but each hydra sectioned at a site differing in the distance from the head and foot of the animal.	1. Regardless at which point a hydra was sectioned transversely, each of the two pieces of the sectioned hydra regenerated into a complete animal. 2. If the section was made in the peduncle (narrowed body tube at the foot end) of a hydra of the third species (H. oligactis), then the regeneration of a new head in that piece was slower.	Only a piece of tentacle was not able to regenerate into a complete hydra.
F. The head was cut from a hydra transversely just below the ringlet of tentacles.	1. The head with ringlet of tentacles regenerated into a small hydra with relatively long tentacles initially. 2. When the small regenerated hydra was fed, it grew into a hydra of normal size.	
G. The head with tentacles described above in I. F. was cut into smaller pieces having two to three tentacles each.	1. Each of the small pieces became a small hydra with two to three relatively long tentacles. 2. When these small hydra were fed, they grew into hydra of normal size, having the normal number of tentacles.	
II. LONGITUDINAL SECTIONS:		
A. Hydra sectioned lengthwise into two pieces.	1. At first each of the two strips coiled, and eventually lengthened. 2. The "outside skin" was on the inside of the coil.	1. Each strip formed a tubular shaped hydra, and eventually regenerated the normal number of tentacles. 2. In forming the tubular shaped hydra, the cut edges on the opposite sides of each strip rejoined along the length of the animal, thereby forming the boundary of a canal that runs the length of the hydra.
B. Hydra was sectioned lengthwise into two strips, and then each of those two strips was sectioned lengthwise again, giving rise to a total of four strips.	Each of the four longitudinal strips of hydra regenerated into complete animals as did those in experiment II. A.	1. It took about three hours for a thin strip to coil, unwind, stretch out, and remain extended. 2. Trembley erroneously concluded that each very thin strip "inflates" itself to form the gut of the regenerated hydra.
C. Hydra quartered by first sectioning it longitudinally into two strips, and then sectioning each strip transversely.	1. Each "head quarter" sealed itself longitudinally, and then regenerated a foot. 2. Each "foot quarter" sealed itself longitudinally, and then regenerated a head.	One of the regenerated hydra was fed, and after it reached full size, it was quartered in a like manner again, and this entire process was repeated one more time.

Table 1. Abraham Trembley's Experiments On Regeneration Of Hydra

	EXPERIMENT	RESULTS	ADDITIONAL OBSERVATIONS
III.	**PARTIAL SECTIONS:**		
A.	Hydra sectioned about halfway down its body beginning at the head end.	The two cut portions healed separately to give a "Y" or "T" shaped hydra having two heads.	1. The experiment was repeated a second time on the two heads that formed to give a four-headed hydra. 2. The four heads were each cut again, and the result was a branched seven-headed "Hydra." 3. The seven heads eventually (weeks later) separated to give seven separate hydra, each with its own head and foot.
B.	Hydra sectioned about halfway up its body beginning at its foot end.	The two cut portions healed separately to give rise to an "inverted Y" shaped animal with one head and two feet.	The two feet of the "inverted Y" shaped animal were cut in the same manner as described in III. B. to give rise to a four-footed hydra.
IV.	**PARTIAL AND RANDOM SECTIONS ON HYDRA OPENED LENGTHWISE:**		
	A hydra was opened by sectioning it lengthwise through only one layer of its "skin" - i.e., one layer of the ectoderm and endoderm - so that a flattened "sheet" of an animal was obtained with the tentacles lined up one by one along the top of the sheet.)		
A.	The sheet was mutilated by many partial sections made along the top, bottom, and two sides of the sheet in such a way that no pieces detached.	A many-headed and many-footed hydra regenerated.	Eventually all of the heads separated as single small hydra.
B.	The sheet was cut by many random sections into many separate small pieces.	1. Many small hydra regenerated from most of the pieces. 2. Some of the pieces did not regenerate and died.	1. When the regenerated small hydra were fed small bits of food, they grew into hydra of normal size. 2. Trembley commented that the pieces that died may have been too small to regenerate.
V.	**SECTIONS MADE THROUGH HYDRA IN THE PROCESS OF BUDDING:**		
A.	Hydra with bud(s) developing was sectioned transversely either above or below the bud(s).	1. The separated head piece regenerated a foot. 2. The separated foot piece regenerated a head. 3. The piece with the bud regenerated a head or a foot depending on which piece was removed.	1. When a head was regenerating on a piece on which a bud was developing, the tentacles regenerated more slowly. 2. The bud always developed regardless of the section made on the parent.
B.	Hydra was sectioned transversely immediately above a developing bud.	The bud remained on the anterior end of the parent's body, formed an obtuse angle with it, and became the mouth of the remaining hydra.	1. Food given to the mouth of the "bud" entered the gut of the parent. 2. Eventually a constriction formed and the end of the former bud separated; the remaining stump subsequently developed tentacles.
C.	A hydra having three buds was sectioned transversely somewhat beneath the head of the parent, and all of the three buds were sectioned in half transversely.	1. The body of the parent lacking a head regenerated one. 2. All of the buds, which were still attached to the parent, regenerated heads. 3. The four pieces having heads, which were separated from the parent and the three buds, all regenerated one foot each.	1. This experiment was conducted on both H. vulgaris and H. oligactis. 2. Subsequently all of the regenerated buds separated from the parent hydra.
D.	Hydra with a bud just starting to develop was sectioned lengthwise.	Each of the two longitudinal strips regenerated as described in II. above.	The bud developing on one of the strips always completes its development and separates from the regenerating strip.

producing a two-headed animal. He repeated the process a number of times and eventually got a seven-headed Hydra. About this monster he writes:

The reader may well imagine that after I had succeeded in making some Hydras, I was not content to stop at that. I cut off the heads of the one that had seven, and a few days later I beheld in it a prodigy hardly inferior to the fabulous Hydra of Lerna. It acquired seven new heads, and, if I had continued to cut them off as they sprouted, no doubt I would have seen others grow. But here is something more than the fable dared invent: the seven heads that I cut from this Hydra, after being fed, became perfect animals; if I so chose, I could turn each of them into a Hydra. (Trembley, 1986, p. 151)

This is the first instance of Trembley's published use of the word "hydra" to refer to any of his polyps. Earlier, in a letter to Réaumur written on December 16, 1741, Trembley had used the French term *Hydre* to refer to a budding hydra (M. Trembley, 1943, p. 70). Linnaeus (1758) later gave this name to the genus we now call *Hydra*.

Buscaglia (1985), in an analysis of the experimental logic of Trembley's regeneration experiments, has pointed out the influence of Trembley's training in infinitesimal calculus. He notes that Trembley treated hydra as cylindrical bodies having an anterior–posterior polarity and that he sectioned them in a geometric pattern, tending to "go to the limits," as one does in calculus.

Most of the experiments and observations described in Table 1 are self explanatory. We should like to comment, however, on a few results that are often overlooked. First, Trembley noted the polarity of regeneration: that the cut edge at the anterior (head) end of a piece of a sectioned hydra developed into a head and that the cut edge at the posterior (foot) end of a cut piece developed into a foot (basal disk). Furthermore, he noted that regeneration was slower in the posterior part of a cut *Hydra oligactis*.

Second, he carefully described the phenomenon of wound healing, especially when reporting the healing of the cut edges of hydra cut in half longitudinally (Fig. 7 in Figure 4.2).

In reuniting, the edges join so well that from the very first moment no scar can be seen in the area where they have joined. The skin of this new polyp is as smooth there as it is elsewhere. Consequently, as soon as the joining of the edges is completed, the bodies of these parts of polyps resemble those of complete polyps . . . [Fig. 8 in Figure 4.2]. The heads bear fewer arms, however, each having only three or four depending on the number of arms possessed by the polyp which was sectioned and depending on how the section was made. (Trembley, 1986, p. 149)

Third, he noted that there was a lower limit to the size of a piece of hydra that could regenerate. Fourth, he noted that if a small piece of sectioned tissue possesses a bud that is beginning to develop, that bud will always complete its development, regardless of the condition of the remainder of the sectioned piece.

It is often erroneously stated that Trembley cut a hydra into fifty pieces and that each of those pieces regenerated into a complete hydra. Trembley obtained the fifty regenerating animals by first cutting a hydra into four pieces. After those four pieces regenerated, he cut each of those four animals into four more, until he eventually obtained a clone of fifty animals from the one with which he had originally started.

Trembley's methods

Trembley carried out most of these experiments holding a hydra in a drop of water in the palm of his hand, using a small scissors to cut with and, at times, a weak magnifying glass to see with. It is instructive to read Trembley's description of his procedures:

To cut a polyp transversely, I place it with a little water in the hollow of my left hand. At first it lies contracted at the bottom of the water. Even when it is in this state it can be cut through easily enough provided one uses a very fine scissors. When the polyp is elongated, however, it is easier to divide it precisely where one wishes. I therefore usually keep the hand containing the specimen still for a moment to allow the polyp time to elongate. When I have it as I want it, I delicately pass one blade of the scissors, which I hold in my right hand, under the part of the polyp's body where it is to be severed. Then I close the scissors, and immediately after having divided the polyp, I examine the two halves under the magnifying glass in order to assess the outcome of the operation. If it has succeeded, I put each of the two portions of the polyp into separate vessels or both together into the same one. One may place the two halves in the same container without fear of confusing them until the posterior part has nearly completed its regeneration.

I placed the parts of the polyps on which I performed my experiments into shallow glass vessels containing only nine to eleven millimeters of water. By this method I could always observe these pieces with a magnifying glass in whatever part of the container they might be.

Since I had a number of vessels at the same time in which I was keeping severed polyps, I marked each one with a number or a letter, using the same symbol in the journal of my observations to identify these polyps. I was the only person to handle the vessels and, when putting in fresh water, I was extremely careful not to mix anything. I took these same precautions with all the polyps upon which I performed the experiments to be reported in this *Memoir*. (Trembley, 1986, p. 142)

How Trembley's discovery became known

Early in his researches, Trembley conveyed the results of his experiments in a letter (December 15, 1741) to the prominent French scientist Réaumur. Through Trembley's cousin, Charles Bonnet, the discoverer of parthenogenesis, Trembley had initiated a correspondence with the renowned French scientist that lasted until Réaumur's death seventeen years later. Réaumur was so excited about Trembley's findings that he not only announced them to the Paris Academy of Sciences but also described them in the preface to the sixth and final volume of his *History of the Insects* (1742).

News quickly spread from the Continent to Great Britain. On November 18, 1742, a letter from J. F. Gronovius, a physician in Leyden, was read to the Royal Society of London and subsequently published in its *Philosophical Transactions* (1744). In it the writer refers to a "Mr. Allemand" as the discoverer of regeneration, "Professors Albinus and Mussenbrock" as verifying the phenomenon, and "Mr. Réaumur" as a potential confirmer, but not to Trembley.

In a footnote to the Gronovius letter, an editor of the *Transactions* refers to "the first account" of the discovery of regeneration as a letter from Georges-Louis Leclerc, comte de Buffon, a prominent French naturalist. In his "first, and perhaps . . . only contribution . . . to the Royal Society [of London]" (Brown, 1946, p. 160), Buffon wrote a letter, dated July 18, 1741, and addressed to the president of the society, Martin Folkes, telling of the regeneration of a complete animal from a piece of one.

The same footnote mentions that at about the same time Folkes had received another letter describing in more detail some of Trembley's findings on regeneration, this one from Charles Bentinck of The Hague, brother of Trembley's employer, William Bentinck, and also brother-in-law of the second duke of Richmond. Bentinck mentioned that the work was done by "a young Gentleman from Geneva, then in Holland." The author of the footnote adds "whose name we since learn to be *Monsieur du Tremblay.*" Although neither the *du* nor the *Tremblay* was correct, the name of Abraham Trembley, the obscure young tutor from Geneva, thus became known to British scientific circles.

Within a short time there followed another seven reports of Trembley's discoveries, which subsequently appeared in volume 42 (1744) of the *Philosophical Transactions,* the volume covering presentations made to the society during 1742–3.

On November 25, 1742, a week after the letter from Gronovius was presented, another paper ("Part of a Letter . . . ," 1744) was read, which summarized Réaumur's account to the Royal Academy of Sciences in Paris.

On December 19, 1742, Folkes wrote directly to William Bentinck, asking him about "the discovery we have heard has been made by a gentleman in your family, of the most remarkable property of an Insect I ever heard of: that of becoming several animals all living . . . upon its being cut into several parts." William Bentinck replied on January 15, 1743; an abstract of that letter appeared in the *Philosophical Transactions* (1744). Trembley's first communication to the Royal Society (1744a) immediately follows Bentinck's note. Trembley's account, translated from the French, includes his first published drawing of a hydra.

Immediately thereafter, on January 21, 1743, there appeared in the *Transactions* a six-page abstract of Réaumur's account (Mortimer, 1744) of Trembley's work, taken from the preface to the sixth volume of Réaumur's *History of the Insects* (1742). Réaumur reported that he and others had repeated Trembley's experiments and confirmed them.

Folkes then initiated a voluminous correspondence with Trembley, copies of which we have obtained from the archives of the Trembley family in Geneva and are currently preparing for publication. Trembley sent Folkes some hydra in March 1743. When he received them, Folkes repeated many of Trembley's experiments and described them in a well-illustrated article in the *Philosophical Transactions* (1744), dated March 24, 1743.

The same volume 42 also includes an article (March 14, 1743) by Charles Bonnet (1744), who not only refers to Trembley's discovery of regeneration but reports that he has himself repeated the same kind of cutting experiment on worms and finds that complete animals also regenerate from cut parts of the worm.

The last reference to Trembley's experiments to appear in volume 42 of the *Philosophical Transactions* is a letter written to Folkes, on May 23/June 4, 1743,[1] by the second duke of Richmond (1744), the man who later was to hire Trembley to raise his eldest son. The letter, read on June 2, 1743, describes in colorful language the experiments the duke witnessed Trembley carrying out while he was a visitor in Holland. Of one experiment the duke writes,

Another Operation I saw him make . . . was that of putting one of the Points of a very small Pair of sharp Scissars into the Mouth of a *Polypus*, and forcing

---

[1] Britain remained on the Julian calendar until after 1750, whereas most European countries had adopted the Gregorian or "New Style" calendar. Thus the British calendar was eleven days ahead of the Continental one.

it out at the very End of the Tail, he then laid it quite open like a *Pigeon*, or a *Barbacute Pig* to be broiled; yet within about Five Hours, I saw the same *Polypus* with the parts so reunited again, that I could not perceive any thing had been done to it. (pp. 510–11)

Another member of the Royal Society, Henry Baker (author of a number of books on the microscope and son-in-law of Daniel Defoe), also a beneficiary of Trembley's generosity in sharing information, jumped the gun a bit and published in 1743, a year before Trembley, his own book on the hydra. He listed the kinds of cutting experiments he had carried out, most of them repeating Trembley's.

Hence, with such a plethora of publications and confirmations in less than one year, it is no wonder that Trembley, an unknown at the outset, was not only elected a fellow of the Royal Society in 1743 but was awarded its coveted Copley Medal the same year. His major opus, *Mémoires, pour servir à l'histoire d'un genre de polypes d'eau douces, à bras en forme de cornes*, was published in 1744. This classic, a book of great beauty and rich content, was translated into German in 1778 by Goeze, into Russian in 1937 by Kanaev, and into English in 1986 by Lenhoff and Lenhoff.

The lively reaction to Trembley's discovery

If asked "Who carried out the first scientific study of regeneration?" one would answer that obviously it was Réaumur. In 1712, when Trembley was only two years old, Réaumur had published his study on the regeneration of the appendages of crayfish. (See Chapter 3 in the present volume.) His work went largely unnoticed, perhaps because he did no more than confirm scientifically what the common folk and fishermen had reported since the times of Aristotle, Pliny, and Augustine. Nonetheless, to answer the question of who carried out the first systematic experiments on regeneration, with regeneration defined as replacing a lost part, the answer is Réaumur.

Trembley's discovery, however, was more remarkable, and far more controversial and unsettling, for he demonstrated the generation of an entire animal from a small piece of that animal. Trembley described a hitherto unknown process by which an animal can generate, a process that required neither egg nor sperm, that did not require copulation. His was thus a finding that called into question the accepted wisdom among his contemporaries on how animals reproduce.

The scientific world was somewhat prepared for Trembley's startling discovery of regeneration, because Charles Bonnet, in 1740, had just described parthenogenesis for the first time. (See Dawson, 1987.) The belief that animals could not reproduce without copulation was further undermined when Trembley reported that hydra reproduce not only by regeneration from cuttings but also by budding asexually, a finding soon confirmed by Trembley (1744b) and others in a number of other animals.

Even though Trembley's evidence for regeneration of complete animals from a small part of one was overwhelming, some of his contemporaries continued to harbor doubts about it because the regeneration experiments had been carried out with a weird animal that few had ever heard of and even fewer had seen. Was it truly an animal or rather a plant, or perhaps a "zoophyte," Leibniz's predicted missing link between the plant and animal kingdoms? (See Josephson, 1985, pp. 348–51.) Voltaire described the hydra – the green hydra, that is – as a vegetable: "This production called a polyp is much more like a carrot or an asparagus than an animal." Others, like Goldsmith, made fun of Trembley's discoveries. (See Baker, 1952, pp. 45–6.)

When Charles Bonnet published his findings showing the regeneration of earthworms (1744), however, more natural historians became convinced that Trembley's findings were neither ambiguous nor an isolated oddity but that regeneration of complete animals from a part of one also occurred in other, better-known creatures. As Charles Dinsmore points out in Chapter 1 of the present volume, Spallanzani widened the list of animals exhibiting the phenomenon of regeneration of lost parts to include amphibians, of the vertebrates.

Theoretical and philosophical reactions to Trembley's discovery

Trembley's discovery that a complete animal can regenerate from a small piece of that animal did not fit well with the views of a large group of natural historians, the preformationists. According to their view of embryonic development, a miniature embryo is present in the egg or sperm, and the embryo develops by the growth of those smaller preexisting parts. This view was supported by such distinguished scientists as Swammerdam and Bonnet.

Trembley's data, however, seemed to support the rival theory of epigenesis, the view that the embryo develops gradually from material that was not preformed. Likewise, materialists like La Mettrie, who believed in the "animal machine," used Trembley's discovery to support their views (Vartanian, 1950). For example, they asked, If an animal were cut into ten pieces and each became a new individual, what did that signify about the original animal soul and its material nature?

From the evidence we have, Trembley himself appears to have avoided becoming involved in arguments about preformation, epigenesis, the animal machine, and the Chain (Scale, Ladder) of Being. Only one brief comment from his later years suggests that despite the support his discoveries seemed to lend to theories of epigenesis, he, as a religious man, leaned toward preformation. (See Baker, 1952, p. 185.) We do not think these kinds of philosophical questions interested him much; he was far more concerned with his findings and with experimental methodology, and he harbored a distaste for theories in general.

Since we are rather like Trembley in this respect, we refer those interested in further exploring these debates to a fine recent book by Virginia Dawson, *Nature's Enigma: The Problem of the Polyp in the Letters of Bonnet, Trembley and Réaumur,* as well as to the writings of Aram Vartanian (1963) and Charles Bodemer (1964).

Bonnet's theory of regeneration

Before concluding this essay with a discussion of Trembley's philosophy of research, a philosophy that we believe to be as meaningful today as it was in his time, we shall pause to consider how Bonnet sought to reconcile his preformationist views and his religious belief in the indivisibility of the soul with the results of Trembley's and his experiments on regeneration. We do this for a special reason: Bonnet's theory of regeneration rather resembles some that a number of contemporary biologists are proposing.

Charles Bonnet (in writings summarized by Kanaev, 1937, and Bodemer, 1964) assumed that the hydra has a "soul" (that is, an organizing principle, not the spiritual concept) in its head region and that beneath the head region the polyp has a number of imperceptible embryos, each of which has its own soul. Thus, Bonnet reasoned, once the head of a hydra (with its soul) is removed, the embryo immediately beneath the removed head becomes active, and its soul becomes the soul of the regenerated hydra. Bonnet thus concluded that when a hydra is sectioned its soul is not divided; rather, conditions are created under which a suppressed soul of one of the invisible embryos is allowed to emerge. If we substitute the phrase "head activator" (Schaller, 1973) for "soul," Bonnet's explanation has a contemporary ring to it.

What today's experimental biologists can learn from Trembley

Trembley made more than fifty original observations and discoveries concerning hydra, which are reported in his *Mémoires* (see Table 1 and Lenhoff and Lenhoff, 1986). Unlike the work of many other eighteenth-century biologists, most of Trembley's discoveries and conclusions still hold today. In fact, the cellular mechanisms underlying many of his fundamental findings, including those of phototaxis, grafting, and asexual reproduction by budding, are currently under study in a number of laboratories and are still almost as much a mystery today as they were when Trembley first discovered these phenomena. The same certainly holds true for regeneration, although, as other authors in the present volume explain, some progress has been made in our understanding of this subject.

If this chapter were to leave you only with the impression that Abraham Trembley, with his discovery and investigation of regeneration, made one of the most significant contributions to experimental biology of the eighteenth century, we should feel that we had failed to present him adequately. We agree with Bonnet, who wrote, more than two hundred years ago:

[Regarding] Trembley's discoveries . . . I almost do not know what to admire more, the miracles of Nature contained in this work [Trembley's *Mémoires*] or the acumen with which they are described. . . . I can recommend this work to all researchers in the natural sciences as the best paradigm of method, out of which they must learn the still too little known art of how to investigate the truths of Nature. (1779–83, vols. 5–6, Art. 207, translation as quoted in Trembley, 1791)

Trembley was a remarkable investigator whose experiments and approach to research helped shape the modern era of experimental biology. An empiricist, wary of relying on preconceived ideas, he stressed the value of experiment, without which, he warned, "instead of clarifying phenomena through new experiments, we have recourse to a hypothesis, or rather a prejudgment, which not only spares us the trouble of observing, but which often serves but to compound our errors" (Trembley, 1986, p. 186). In addition to being an empiricist who insisted on postponing the development of theory until he had sufficient data, he was an observer quick to see the unusual and able to report his findings with great accuracy and detail. He was an experimentalist who was not content until he could prove his findings a number of ways. He used a quantitative approach, backing up many of his experiments with numbers and repeating his experiments until he was convinced of their validity.

Thus, one should not become disheartened by want of success, but should try anew whatever has failed. It is even good to repeat successful experiments a number of times. All that it is possible to see is not discovered, and often cannot be discovered, the first time. (Ibid., p. 104)

Trembley was what we might call an "operationalist," for he believed an experiment has no lasting value unless the methodology is described in a way that enables others to replicate it.

It is not enough to say, therefore, that one has seen such and such a thing. This amounts to saying nothing unless at the same time the observer indicates how it was seen, and unless he puts his readers in a position to evaluate the manner in which the reported facts were observed. (Ibid., p. 2)

Trembley was organismic in approach, interested not in a single problem, but in all that an animal could teach him. He validated the importance of basing the study of living organisms on direct, careful observation, rather than on preconceived ideas and theory. He believed that it was only by doubt, observation, experiment, and openness to chance discoveries that he had been able to reveal the phenomenon of regeneration.

In order to extend our knowledge of natural history, we must put our efforts into discovering as many facts as possible. If we knew all the facts that Nature holds, we would have the explanation of them; we would see the whole which together these facts fashion. The more we know of them, the more we will be in a position to delve deeply into some parts of this Whole. Thus we cannot work better to explain the facts we know than by trying to discover new ones. Nature must be explained by Nature and not by our own views. (Ibid., p. 187)

ACKNOWLEDGMENTS

We thank the Trustees and Faculty of Jesus College, Oxford University, for serving as gracious hosts to Howard M. Lenhoff while he was a visiting senior research fellow, in Trinity Term, 1988; the Bodleian Library, Oxford; the British Museum, London; and the library of the Royal Society of London. We also thank the family of the late Jean-Gustave Trembley, Geneva, for allowing us to delve into the family archives, particularly into their collection of the M. Folkes – A. Trembley correspondence.

REFERENCES

Baker, Henry. 1743. *An Attempt towards a Natural History of the Polype.* R. Dodsley, London.

Baker, John R. 1952. *Abraham Trembley of Geneva: Scientist and Philosopher, 1710–1784.* Edward Arnold, London.

Bentinck, William. 1744. Abstract of part of a letter from the Honourable William Bentinck, Esq; F.R.S. to Martin Folkes, Esq. *Philos. Trans. R. Soc. Lond.* 42: ii [282].

Bodemer, Charles W. 1964. Regeneration and the decline of preformationism in eighteenth-century embryology. *Bull. Hist. Med.* 38: 20–31.

Bonnet, Charles. 1744. On Insects which are multiplied, as it were, by Cuttings or Slips. *Philos. Trans. R. Soc. Lond.* 42: 468–88.

—— 1779–83. *Oeuvres d'histoire naturelle et de philosophie.* 18 vols. S. Fauche, Neuchâtel.

Brown, Harcourt. 1946. *Buffon and the Royal Society of London.* In M. F. Ashley Montagu, ed., *Studies and Essays in the History of Science and Learning: Offered in Homage to George Sarton on the Occasion of His Sixtieth Birthday,* pp. 141–65. Schuman, New York.

Buscaglia, Marino. 1985. The rhetoric and proof utilized by Abraham Trembley. In H. M. Lenhoff and P. Tardent, eds., *From Trembley's Polyps to New Directions in Research on Hydra. Arch. Sci.* 38 (3): 305–19.

Dawson, V. P. 1987. *Nature's Enigma: The Problem of the Polyp in the Letters of Bonnet, Trembley, and Réaumur.* American Philosophical Society, Memoir 174. Philadelphia.

Folkes, M. 1744. Some account of the insect called the fresh-water polypus, before-mentioned in these Transactions. *Philos. Trans. R. Soc. Lond.* 42: 422–36.

Folkes, M., to W. Bentinck. December 19, 1742. Archives of the Jean-Gustave Trembley Family, Geneva.

Goeze, J. A. E., see A. Trembley, 1791.

Gronovius, J. F. 1744. Extract of a letter from J. F. Gronovius, M. D. at Leyden, November 1742, to Peter Collinson, F.R.S. concerning a water insect, which, being cut into several pieces, becomes so many perfect animals. *Philos. Trans. R. Soc. Lond.* 42: 218–20.

Josephson, Robert K. 1985. Old and new perspectives on the behavior of hydra. In H. M. Lenhoff and P. Tardent, eds., *From Trembley's Polyps to New Directions in Research on Hydra. Arch. Sci.* 38 (3): 347–58.

Kanaev, I. I., see A. Trembley, 1937.

Leeuwenhoek, Antony van. 1704. Concerning green weeds growing in water, and some animalcula found about them. *Philos. Trans. R. Soc. Lond.* 23: 1304–11.

Lenhoff, Sylvia G., and Howard M. Lenhoff. 1986. *Hydra and the Birth of Experimental Biology, 1744: Abraham Trembley's Memoirs Concerning the Natural History of a Type of Freshwater Polyp with Arms Shaped Like Horns.* Boxwood Press, Pacific Grove, Calif. [See also A. Trembley, 1986.]

Linnaei, Caroli. [Linnaeus]. 1758. *Systema naturae,* ed. 10, vol. 1. Reprinted., 1956, *Systema naturae: A Photographic Facsimili.* British Museum of Natural History, London.

Mortimer, C., ed. 1744. An Abstract of what is contained in the Preface to the Sixth Volume of Mons. Réaumur's History of Insects, relating to the above-mentioned Observations, and delivered in to the ROYAL SOCIETY, immediately after the foregoing Paper. *Philos. Trans. R. Soc. Lond.* 42: xii–xvii [292–7].

Part of a Letter from _____ of Cambridge, to a Friend of the ROYAL SOCIETY, occasioned by what has lately been reported concerning the Insect mentioned in Page 218 of this Transaction. 1744. *Philos. Trans. R. Soc. Lond.* 42: 227–34.

Réaumur, René-Antoine Ferchault de. 1742. *Mémoires pour servir à l'histoire des insectes.* Vol. 6. Imprimerie Royale, Paris.

Richmond, Lennox, and duke of Aubigné. 1744. Part of a letter from his Grace the Duke of Richmond, Lennox and Aubigne, F.R.S. to M. Folkes, Esq. *Philos. Trans. R. Soc. Lond.* 42: 510–13.

Schaller, H. C. 1973. Isolation and characterisation of a low molecular weight substance activating head and bud formation in *Hydra. J. Embryol. Exp. Morphol.* 29: 27–38.

Trembley, Abraham. 1730. *Thesis mathematicae de infinito et calculo infinitesimali.* M. Bousquet, Geneva.

    1744a. Observations and experiments upon the fresh-water polypus. (Translated from the French by P.H.Z.) *Philos. Trans. R. Soc. Lond.* 42: iii–xi [283–291].

    1744b. *Mémoires, pour servir à l'histoire d'un genre de polypes d'eau douce, à bras en forme de cornes.* Verbeek, Leiden.

    1791. *Des Herrn Trembley Abhandlungen zur Geschichte einer Polypenart des süssen Wassers mit hörnerförmigen Armen.* (Translated from the French by J. A. E. Goeze.) Reussners, Quedlinburg. [First published 1775.]

    1937. *Mémoires, pour servir à l'histoire d'un genre de polypes d'eau douce, à bras en forme de cornes.* (1744). (Translated into Russian by I. I. Kanaev.) State Biological and Medical Press, Moscow.

    1986. *Mémoires, pour servir à l'histoire d'un genre de polypes d'eau douce, à bras en forme de cornes.* (1744). (Translated by S. G. Lenhoff and H. M. Lenhoff.) Boxwood Press, Pacific Grove, Calif. See Lenhoff and Lenhoff, 1986.

Trembley, Maurice, ed. 1943. *Correspondance inédite entre Réaumur et Abraham Trembley.* Georg, Geneva.

Two letters from a gentleman in the country, relating to Mr. Leuwenhoeck's letter in Transaction no. 283. 1704. *Phil. Trans.* 23: 1494–1501.

Vartanian, Aram. 1950. Trembley's polyp, La Mettrie, and eighteenth-century French materialism. *J. Hist. Ideas* 11: 259–86.

    1963. *Diderot and Descartes: A Study of Scientific Naturalism in the Enlightenment.* Princeton University Press, Princeton.

5

Lazzaro Spallanzani: concepts of generation and regeneration

CHARLES E. DINSMORE

In 1768, Lazzaro Spallanzani (1729–1799) published his *Prodromo di un opera da imprimersi sopra la riproduzioni animali,* a preliminary report of his studies on animal reproduction. In it he provided a broad outline of his regeneration research, confirming previous studies but dramatically extending the range of inquiry to include vertebrates. Though perhaps not one of his best-known publications, it discloses a core of experimental studies that influenced the development of subsequent ideas on generation. The *Prodromo* was intended, judging from its title and its table of contents, as a preliminary essay on regeneration, setting out the parameters of a more comprehensive work on animal reproduction that was to follow. Unfortunately, the latter work was never written.

Another objective, perhaps the principal one, was to gain Spallanzani membership in the Royal Society of London. The manuscript was sent to the Royal Society as a "present" and as an example of his research. Its translation into English the following year by the society's secretary, Matthew Maty, assured the work a more extensive readership. And with letters of recommendation from Charles Bonnet (1720–93) and Abraham Trembley (1710–84), themselves members of the society, Spallanzani was subsequently elected a foreign correspondent of the society.

Most students of regeneration are aware of some of the studies Spallanzani mentioned in the *Prodromo*. Writing from extensive personal observation, he included brief chapters on the regenerative ability of earthworms, aquatic boat worms, slugs and snails, legs of aquatic salamanders, tails of aquatic salamanders, tails of tadpoles, jaws of aquatic salamanders, and legs of frogs and young toads. After the chapter on tadpole tail regeneration he inserted a treatise that, though not about regeneration, reveals Spallanzani's natural philosophy. It is entitled "Of the Existence of the Tadpoles in the Eggs before Fecundation." Spallanzani was an ovist preformationist. We shall explore the basis for this as we examine his intellectual development.

67

The scope of this essay is modest compared with the wide range of the man's intellect and career; he published studies on plant and animal physiology that described his many creative experiments on digestion, respiration, and artificial insemination.[1] He was also actively interested in geology and volcanology.[2] This essay focuses on Spallanzani's role in the controversies surrounding the nature of "generation," particularly his use of regeneration research as a means of exploring *questions ténébreuses* – questions that went to the heart of deep philosophical issues of his day.

One element of the intellectual context that I wish to examine is the nature of "generation" as it was understood in the early eighteenth century by those engaged in the preformation–epigenesis debate. (For an excellent analysis of this issue, see Roe, 1981.) I highlight the epigenesis–preformation dispute as it was argued between John Turberville Needham (1713–81) and Charles Bonnet, to whom the idea of spontaneous generation was a bête noire. Their comments show why regeneration became a critical biological, as well as philosophical, issue during this period.

Another major influence on Spallanzani's intellectual development was his elaborate correspondence with Bonnet. From their letters one can begin to grasp the intensity that each man brought to the problem of accommodating new observations on "generation" within a reasonable – that is, philosophically acceptable and internally consistent – contemporary worldview. Although Bonnet had the more profound understanding of contemporary philosophical issues and their theological implications, it was Spallanzani's energy that carried the relationship and produced real scientific progress. I shall be considering for the most part only the first five years of their correspondence, when regeneration research was a common topic.

Finally, it is important to recognize the impact that Spallanzani's experimental studies on animal regeneration had on eighteenth-century biological thought. Considering the technical limitations of his studies, it is a tribute to his observational skills to note how modern some of his findings appear. Before examining these issues, however, we must establish an understanding of Spallanzani as an individual.

[1] For example, his two-volume *Dissertazioni di fisica animale e vegetabile*, published in Modena in 1780, describes his experiments on digestion and reproduction, including artificial insemination. It was soon translated into both French (by Senebier, 1786) and English (by Beddoes, 1784).

[2] His multivolume *Viaggi alle due Sicilie* . . . (Pavia, 1792–7) includes personal accounts of geological observations that he made during his travels, as well as amusing anecdotes about his encounters with volcanoes.

Lazzaro Spallanzani: a brief biography

Born in 1729 to a large, well-to-do family in the small town of Scandiano in northern Italy, Lazzaro attended local schools until the age of fifteen. He was then enrolled in the Jesuit seminary in nearby Reggio Emilia, where his academic skills began to flower, especially in rhetoric, philosophy, and languages. His intellectual ability appears to have earned him the nickname *l'astrologue* (the astrologer) among the students (Rostand, 1951, p. 9). In 1749 he matriculated at the University of Bologna, and he worked for three years toward a doctorate in jurisprudence. There he came under the influence of his father's cousin, Laura Bassi, who was professor of mathematics and physics. Finding mathematics and the natural sciences more to his liking than the law, with the help of Bassi and another professor, Antonio Vallisneri the younger, he secured the consent of his father to abandon jurisprudence.

By 1754, Spallanzani had earned his doctorate in philosophy. Having also received instruction in metaphysics and theology, he took minor orders, was subsequently ordained as a Roman Catholic priest, and became attached to two congregations in Modena. According to Dolman (1975), his "casual religious commitments" were performed irregularly, and he devoted himself to intellectual pursuits in the natural sciences. His ties with the church however, provided him with financial support. (He appears to have had no private income.) It also offered moral protection against the Italian Inquisition, which still censored research and publications it considered contrary to Catholic doctrine. Publications in natural history were suspect, so Spallanzani's clerical role "facilitated his investigations of natural phenomena" (Dolman, p. 553). A series of academic appointments began in 1755 at the College of Reggio Emilia, where he mostly taught humanities. He made it clear, however, in a letter to Antonio Vallisneri, written in October 1761 (Biagi, 1958, vol. 1, p. 24), that he was not happy and wanted to obtain an appointment at a university. In 1763 he accepted an invitation to become professor of philosophy at Modena, in both the university and the College of Nobles, but again he was not entirely happy with his teaching responsibilities, writing to Bonnet, on April 17, 1766, "I am not the master of my own time. The physics [i.e., natural philosophy] and mathematics that I must teach rob me of much of it" (ibid., p. 84).[3]

[3] "Je ne suis pas tout à fait maître de mon tems. La Phisique et la Mathematique que je dois enseigner m'en dérobent beaucoup." No attempt has been made to edit Spallanzani's French. (Translations from the French throughout this chapter are my own.)

During this period, he consumed volumes on the natural sciences, devouring works by or about Newton and Leibniz as well as many other contemporary natural philosophers. A characteristic that appears early in his correspondence, however, is that while he was intensely interested in the works of his illustrious predecessors as well as in the opinions of his esteemed contemporaries, he was a quintessential skeptic, reluctant to accept anything he read that he did not confirm with his own hand. He took full advantage of the library and extensive network of contacts of his professor Antonio Vallisneri in procuring the books that he wanted before setting about his experiments. The collected correspondence shows many requests by Spallanzani for the loan of books, as well as comments on them and expressions of gratitude for Vallisneri's generosity (e.g., Biagi, 1958, vol. 1, pp. 19–20). In this way he was introduced to, and became fascinated by the ideas and experiments on generation that he found in works by Needham and by the great French naturalist Georges-Louis Leclerc, comte de Buffon (1708–88). Early in 1761 (at the end of *Carnival*), he asked Vallisneri for Buffon's work on generation (presumably the *Histoire naturelle*). He wanted to read it again before attempting to replicate some of the experiments that it described. He wrote soon after, on April 6, 1761 (ibid., p. 19), to share with Vallisneri some of his thoughts arising from preliminary observations on animalcules that he had seen in his infusions. He was clearly disturbed by the philosophical implications of Buffon's and Needham's work, which, if verified, supported the theory of spontaneous generation. For this reason, he suggested that their studies ought to be tested more rigorously. A man of no small ego, he appeared to think nobody better qualified to do this than himself.

Spallanzani then obtained a better microscope and wrote to Vallisneri, on June 1, 1761 (Biagi, 1958, vol. 1, p. 21), that he had seen the "little moving bodies" (*i corpicciuoli guizzanti*) just as Needham had described them. He was at that point in complete agreement with Needham's observations and, indeed, told Needham so (Castellani, 1971, p. 2). Nevertheless, he subsequently undertook an extensive series of experiments in which he carefully repeated Needham's studies on the genesis of animalcules in plant and animal infusions. Utilizing his own data and observations, Spallanzani proclaimed his entrance into the field of generation by publishing his *Saggio di osservazione microscopiche concernenti il sistema della generazione de Needham e Buffon* (1765). In it he demolished the studies of Needham and Buffon, showing their techniques to have been inadequate and their conclusions favoring epigenesis or spontaneous generation completely unwarranted. In that same year he began his lifelong correspondence with Charles Bonnet and, under Bonnet's influence, initiated his studies of regenerative phenomena.

At last, in 1769, a year after he had published his *Prodromo,* he accepted an offer to be professor of natural history at the University of Pavia, a position for which he had been preparing all of his life and in which he maintained a physically vigorous and academically productive career for the next thirty years.

This has been a broad outline of some of the events in Spallanzani's career that help us to understand his contributions to the history of regeneration research. Before examining specific questions and issues that he pursued so zealously, note must be taken of the contemporary philosophical context that led him to his investigations and interpretations of generation and its surrogate, regeneration.

Concepts influencing natural history in the eighteenth century

Among the eighteenth-century concepts influenced by earlier progress in physics, mechanics, and mathematics, attributable in large part to the dissemination of the works of Descartes, Newton, and Leibniz, was the issue of the nature of generation. Moreover, this appears to have been one of the main issues debated among European natural historians and philosophers in the late seventeenth and throughout the eighteenth century. It is also illustrative of "the fundamental tie between biological and philosophical questions that existed in the Enlightenment period" (Roe, 1981, p. 2).

Epigenesis and preformation

The competing theories of epigenesis and preformation created a dynamic tension concerning how biological generation was to be understood. Epigenesis, in particular, was thought to be a variation on the theologically disturbing concept of spontaneous generation. As articulated most clearly in the writings of Buffon and Needham, the concept of epigenesis relied on a theoretical framework of undifferentiated "organic molecules" and internal molds that received and organized them.[4] With epigenesis, the role of God in generation and development was open to question. Preformation, however, with its

---

[4] For an excellent discussion of the philosophical ramifications of this concept, see Roe's (1981) analysis of the Haller–Wolff controversy.

assumption that all generations of creatures were established one within the other at the time of Creation, found favor because of the ease with which it fit theologically based philosophical needs. Thus a constraining component of seventeenth- and early eighteenth-century biological thought was the lingering belief that there were inviolable natural laws, which could be gleaned from the study of books written by acknowledged historical authorities. These included, among other sources, Aristotle, the Judeo-Christian Bible, Galen, and Pliny. But the conceptual revolutions in physics and mathematics that brightened the intellectual horizons of the seventeenth century began to illuminate some of the flaws in the contemporary paradigm related to studies of living things. Among the most interesting and least understood phenomena was reproduction or, in the generic sense, generation.

In the unstable philosophical atmosphere created by the competing theories of generation, the stage was set for a paradigm shift, though one of minor proportions when compared with that engendered by Newton's works. Nevertheless, it seems clear that the new horizons created by animal regeneration experiments had a revolutionary effect on the eighteenth-century concept of generation. Moreover, Spallanzani carried regeneration research to vertebrates; beyond the safe speculations about the implications for "insects," a broadly defined category of lower animals, he raised the question of its application to, and significance for, man.

Exploration of the fundamental nature of generation also became a focus of the new empirical and experimental approach to nature. People began to pose to living Nature the rudely confrontational question "What if . . . ?" and, in so doing, began the modern era of experimental biology. But how was Spallanzani introduced to these issues, and how did he become involved in their debate?

Spontaneous generation and regeneration

To place Spallanzani's regeneration research in the appropriate context, we need to reflect briefly on the history of the concept of spontaneous generation. In 1668, Francesco Redi had published a summary of his pivotal studies disproving spontaneous generation with his observation that eggs laid by flies accounted for the origin of maggots in decaying meat. Although they satisfied the standards for the larger, visible world, Redi's experiments were not sufficiently conclusive to extinguish the belief that the newly discovered microscopic worlds might operate under different guidelines that could include spontaneous generation. The final proof had to await the experiments of Louis Pasteur, who, in the nineteenth century, paid tribute to Spallanzani, saying that the latter had had *le reflexe expérimental* (Rostand, 1951, p. 29). Spallanzani, it should be noted, had twice referred to Redi's work in the *Prodromo*, published a century after Redi's famous opus. Spallanzani himself thus made a direct connection between the study of regeneration and the concept of generation.

In Spallanzani's day, however, the principle of spontaneous generation was reborn under the rubric of epigenesis and championed by Needham and Buffon, as noted earlier. Spallanzani demonstrated some familiarity with the studies of both men in his correspondence with Vallisneri (e.g., April 6, 1761; June 1, 1761; Biagi, 1958, vol. 1, pp. 19–22). Repeating Needham's experiments on animal and plant infusions, he communicated his preliminary findings to Needham, indicating that they were in accord with his epigenetic interpretation of generation. Needham, in turn, gladly transmitted this impression in a letter on February 13, 1762, to Bonnet (Castellani, 1971, p. 2): "A professor from Reggio has just written to me that he has made precisely the same observations, to which he has added many others, to confirm my previous feelings [about an epigenetic interpretation of generation]."[5]

Bonnet was delighted that Needham later, though temporarily, modified his stance, after Spallanzani published his completed study (i.e., the *Saggio*, as noted earlier). In his first letter to Spallanzani, on September 14, 1765, Bonnet congratulated the younger Italian, telling him that he had "succeeded in removing the blindfold that has covered our learned Colleague's [Needham's] eyes. What I had only initiated, you have finished, and what I had only glimpsed, you have seen." (Castellani, 1971, p. 3).[6] Nevertheless, Needham maintained his epigenetic perspective, which Bonnet and Spallanzani referred to in their subsequent correspondence as a bizarre philosophy.

Parthenogenesis, budding, and regeneration

Spallanzani's perception of natural history and, in particular, his understanding of the nature of generation were strongly shaped by

---

[5] " . . . un professeur de Reggio me vient d'écrire qu'il à fait précisément les mêmes observations, à lesquelles il a ajouté plusieures autres, pour confirmer mes sentiments la-dessus."

[6] "C'est ainsi, Monsieur, que vous avez réussi à enlever le bandeau qui couvrait les yeux de notre savant Confrère. Ce que je n'avais qu'ébouché, vous l'avez fini, et ce que je n'avais qu'entrevu, vous l'avez vu."

other issues besides spontaneous generation. Toward the middle of the eighteenth century, as Dawson (1984, p. 43) has noted, two powerful discoveries revolutionized ideas about the potential processes and mechanisms of animal reproduction. One was Trembley's 1740 discovery (published 1744; trans. Lenhoff and Lenhoff, 1986) not only that polyps or hydra could regenerate missing parts but, more important, that they were also able to produce more of their own kind, either by budding or from cuttings. (See also Chapter 4 in the present volume.)

The other discovery was Bonnet's confirmation of the previously anticipated capacity of some lower forms to undergo asexual reproduction, or parthenogenesis. It solidified Bonnet's influential opinion in favor of ovist preformation, a position Spallanzani subsequently shared. (See also Hankins, 1985, relating these issues to the Enlightenment in general.)

Beyond the natural history context, what did these generation-related discoveries mean for contemporary intellectuals? Because Spallanzani developed a close relationship with Bonnet, who strongly influenced his opinions and guided much of his development as an investigator, we need to explore for a moment Bonnet's earlier philosophical approach to this topic, as expressed in his letters. A letter dated June 29, 1741, from Bonnet to one of his former teachers, regarding Trembley's discovery, is most enlightening on this point. Writing about the philosophical implications of polyp regeneration, Bonnet was not concerned with the issue of "germs" as opposed to eggs, a thesis that he championed in his scientific discussions. Instead he demands of his professor: "Shall we allow it [the polyp] a soul or not?" (quoted in Dawson, 1987, p. 141). A few months later (September 1, 1741) he wrote to Trembley, asking several metaphysical questions relating to the soul. Among them: "Are there in this Insect . . . as many souls as there are portions of these same Insects which can themselves become perfect Insects?" (ibid., p. 162). It is clear also from his later writings and correspondence that this was an overriding concern. As Rostand (1951, p. 40) observed, Bonnet eventually developed a metaphysical perspective on generation bordering on mysticism. On what was this preoccupation based?

A Leibnizian connection

This philosophical concern had its origin in the influential writings of Gottfried Wilhelm von Leibniz (1646–1716), among other thinkers, whose works informed the dominant opinions of that era. In 1695, Leibniz described his New System of Nature, expanding on his ideas regarding the union or unity of the soul and body. "I saw that these forms and these souls should be indivisible, . . . that that was the thought of Saint Thomas regarding the souls of animals" (Wiener, 1951, p. 108). That animals had souls was for Leibniz apparently a given assumption. The behavior of the bisected polyp, not to mention that of the higher forms on which Spallanzani subsequently experimented, caused obvious difficulties for this theory.

In a subsequent essay (Wiener, 1951, p. 195), Leibniz included the comment "I believe that not only the soul but also the animal itself subsists." He continued: "It may be doubted whether an altogether new animal is ever produced, and whether animals wholly alive as well as plants are not already in miniature germs before conception." And, furthermore, "the laws of mechanics alone could not form an animal where there is nothing yet organized." He therefore concluded with "the fact that . . . there is mechanism in the parts of the natural mechanism *ad infinitum,* and so many folded one within another, that an organic body never could be produced altogether new and without any preformation." This passage is particularly instructive because it was later transcribed in full by Bonnet in a letter to Spallanzani, on January 17, 1771 (Castellani, 1971, pp. 188–9). In it we find both the fundamentals of preformation and the *emboitement* concept so often attributed to Bonnet. Here then one discovers at least the immediate source of Bonnet's and Spallanzani's ideas on the soul in generation, as well as their concept of preformation based on the *emboitement* of germs. It was immediately apparent that the ramifications for regenerative phenomena presented major difficulties for their paradigm.

Plant–animal continuity, or the Chain of Being

Leibniz's opinions on the relationship of plants, animals, and nonliving materials provided another domain for intense philosophical debate. In this arena too, Spallanzani's opinions developed under the influence of Leibniz's philosophy and writings, as can be seen in some of Spallanzani's early correspondence with Vallisneri.

The essential characteristics of Leibniz's Chain of Being were plenitude, linear gradation, and continuity; the last two of these figured prominently in Spallanzani's interpretation of regeneration. In a letter of 1702, Leibniz, describing his concept of the gradation of the different classes of beings, compared them to "so many ordinates of the same curve whose unity does not allow us to place some other ordinates between two of them." His concept of continuity made him willing to accept the idea that some beings might share the characteristics of both plants and animals: "The existence of Zoophytes, or as Buddaeus calls them Plant-Animals, is nothing freakish, but it is even befitting the order of nature that there should be such." He speculated further that, "creatures might be discovered which . . . could pass for either vegetables or animals. . . . I am even greatly persuaded that there ought to be such beings" (Wiener, 1951, p. 187).

How does this correlate with regeneration research? Trembley's dissertation at the Academy of Calvin in 1731 was on the infinitesimal calculus, discovered by Newton and Leibniz. For Leibniz, the calculus "provided a striking mathematical justification" (Dawson, 1984, p. 45) in support of his concept of biological continuity, a concept of which Trembley was well aware, and with which he grappled in his thinking about the polyps. Trembley himself rejected the designation "plant-animal" or "zoophyte" as inappropriate until all of the attributes of both plants and animals had been more precisely defined and understood (Dawson, 1987, p. 103).

The connection between Leibnizian concepts and Spallanzani's philosophical reasoning on continuity and preformation is found not only in his relationship to the Genevans, Trembley and Bonnet, but also in his close association with Vallisneri. In a 1714 letter, Leibniz stated, "I wish indeed that there were a more thorough investigation of the big question of the generation of animals which should be analogous to that of plants. . . . Nobody is more qualified to clear up this doubt than Vallisnieri [sic] and I hope very much to see his dissertation soon" (Wiener, 1951, p. 199). Leibniz here refers to the father of the professor to whom Spallanzani owed much of his early exposure to science. Indeed, the younger Vallisneri succeeded his father at Bologna in the chair of theoretical medicine and edited his father's collected works, which included research on animal reproduction (Montalenti, 1971).

These, then, were some of the powerful and ubiquitous philosophical issues that informed the opinions of eighteenth-century European intellectuals. They were frankly debated in the circle of correspondents to which Spallanzani belonged and were thus a part of the context supporting his interpretations of generation and regeneration. A direct examination of letters from the elaborate correspondence between Spallanzani and Bonnet will permit us to examine their relationship in some detail.

The correspondence between Spallanzani and Bonnet

The early letters

In several early letters to Vallisneri, Spallanzani expressed great admiration for Bonnet as an outstanding philosopher and preeminent natural historian. He subsequently initiated correspondence with Bonnet, on July 18, 1765, sending him copies of two of his own publications on natural history: one in Latin on the mechanics of skipping stones on water (*De labidibus ab acqua resilentibus*) and the other in Italian (the *Saggio di osservatione microscopiche . . .* that was noted earlier), known collectively as the *Dissertazione due* (1765). In the accompanying brief cover letter, he solicited Bonnet's comments on his Italian dissertation, "which revolves around a point that has much in common with your very famous work on organized bodies [organisms]" (Biagi, 1958, vol. 1, p. 55).[7] He sent the same packet to Albrecht von Haller (1708–77) on the same day and with only slight modifications in the cover letter (ibid., p. 56).

A month later, on August 24, Spallanzani wrote again to Bonnet, though in Italian this time (ibid., p. 63), with a more explicit critical comment on Needham and Buffon's "System of Nature" as a supplement to the *Saggio* he had sent earlier. Bonnet's response was delayed, however, owing to his inability to read Italian; the contents of both the *Saggio* and the letter thus were not immediately accessible to him. Therefore, his recognition of Spallanzani as an enthusiastic proponent of his own views was postponed until his good friend and cousin Abraham Trembley could provide him with a translation. Bonnet was subsequently pleased to discover that Spallanzani had read his books on natural philosophy and supported the ideas contained in them. Moreover, he was particularly delighted to learn that Spallanzani had begun studying earthworm regeneration, an investigation that Bonnet promoted in his *Corps organisés* (1762). To this introduction from Spallanzani, Bonnet responded on September 14, 1765.

Bonnet had previously heard about Spallanzani from Needham. Needham had written to Bonnet that he confidently expected Spallanzani's work, when published, to support his own epigenetic interpretation of the origin of "animalcules" (transcribed in Castellani, 1971, p. 2). (Bonnet, however, was a staunch preformationist, strongly opposed to the theory of epigenesis.) A paragraph in Bonnet's Sep-

---

[7] ". . . roule sur un point, qui a tant de rapport avec votre très célèbre ouvrage sur les corps organizés."

tember 1765 letter to Spallanzani shows clearly that he enjoyed Needham's humbling after the publication of Spallanzani's *Saggio*. This
early exchange established a firm rapport between the two men, based
on (among other things) their shared disdain for epigenetic theories
of generation. Bonnet's theoretical speculations would now merge
with Spallanzani's brilliant experimental studies, creating not only a
powerful bond between the two but also a formidable paladin for
preformation.

Bonnet closed his first letter to Spallanzani by saying that he was
having a copy of his two-volume book *Contemplations de la nature* sent
"as a small token of the esteem that you have inspired in me" (Castellani, p. 5) He expressed his particular pleasure in discovering that
Spallanzani had taken up his regeneration challenge and begun working on earthworms: "I am very pleased with the invitation that I've
made to naturalists about training themselves by studying earthworms, since it has induced you to pursue their study."[8] Bonnet then
charged Spallanzani directly with carrying forward the important
work on worm regeneration: "Persevere in giving them the attention
that they deserve: they will surely repay you for your troubles and
your patience."[9] He then added, "I am eagerly awaiting your thought-
ful studies; they interest me very strongly" (Castellani, pp. 5–6).[10]
How could Spallanzani refuse this ultimate invitation? The impression
was sufficient to warrant mention in Spallanzani's *Prodromo* three
years later. A link between Spallanzani's early experiments on generation, his correspondence with Bonnet, and his entrance into the realm
of regeneration research is thus established.

In his next letter to Bonnet, dated November 18, 1765, Spallanzani
outlined a number of experiments that Bonnet's writings had inspired. These included sectioning not only "thousands" of earthworms
and other "insects" but also amputating the tails of salamanders. "I
haven't overlooked the aquatic salamanders. I observed that the tail
of these animals after amputation completely reproduces itself." This
discovery captured his imagination, and he declared, "To me, this
phenomenon appears appropriate for the clarification of many oth-

---

[8] "Je me sais bon gré de l'invitation que je faisais aux naturalistes de s'exercer
sur les verres de terre, puisqu'elle vous a déterminé à les observer." Trembley
had communicated his hydra regeneration discovery to his friend and cousin
Bonnet within months of his first observations. By June of 1741, Bonnet had
begun his own regeneration research on worms and wrote of his findings in
several publications, both scientific and philosophical.

[9] "Persévérez à leur donner l'attention qu'ils méritent: ils récompenseront sûre-
ment vos peines et votre patience."

[10] "J'attends beaucoup vos savants recherches, et elles m'intéressent fort."

ers,"[11] referring here to the problems associated with animal "reproductions" discussed in both Bonnet's and Needham's speculative writings. He added that "the new tail [regenerate] of the salamanders . . . is very transparent" and that its internal organization could be readily observed with a microscope (Biagi, 1958, vol. 1, pp. 68–9). He could not have helped noticing the circulating blood in these transparent tail regenerates, an observation that likely planted the seeds for his subsequent studies in circulatory physiology.

Of significance to our understanding of the philosophical framework within which Spallanzani was developing his research orientation is his comment in the following paragraph. "I would have, in my imagination, passed from the Animal Kingdom to that of Plants, making the same experiments on each, and comparing the cuts and regenerates of the animals with those of the plants" (Biagi, vol. 1, p. 69).[12] This simple declaration is an explicit reference to the concept of continuity between plants and animals about which Leibniz had speculated.

Bonnet responded to Spallanzani's preliminary research notes and flattering comments on December 27, 1765, with an elaborate letter (Castellani, pp. 6–15). He discussed specific issues raised in the *Saggio,* which Trembley had translated for him, and then praised Spallanzani's conclusion that "animalcules" in infusions arise from eggs: "Generation by ambiguous means must be banished from our philosophy."[13] So much for Needham, Buffon, and epigenesis. Once again, he exhorted Spallanzani "to continue your experiments on earthworms" and offered several reasons why they are the best animals for the study of regeneration.

Addressing Spallanzani's observations on salamander regeneration, he expressed great pleasure in Spallanzani's discoveries. In addition, he suggested specific events to monitor more closely, since they might provide key information relative to the mechanisms of generation. He also mentioned the "germ" concept as one of the problems that regeneration research might help to resolve. Seeing regeneration as a means of disproving epigenesis, he stated: "We will not assume that a few molecules brought together by chance can produce a tail or any other

---

[11] "Les salamandres aquatiques n'en ont pas été oubliées. J'ai observé que la queue de ces derniers animaux après la coupure s'est reproduite en entier. Ce phénomène me paroit à propos pour en éclaircir beaucoup d'autres."

[12] "J'aurois dans la phantasie de passer du Regne animal au végétal, faisant les mêmes expériences sur celui-ci, et comparant les coupures et les reproductions des animaux, avec celles des végétaux."

[13] "Les générations équivoques doivent être bannies de notre philosophie."

part. This way of thinking would be the death of philosophy" (Castellani, p. 10).[14] Later the same letter, he reinforced the antiepigenesis attitude, claiming:

A false philosophy is trying its best to present these animalcules to us as bastards of Nature [an allusion to Buffon and Needham and their theory of epigenesis]: . . . multiply as much as you can the proofs of their legitimacy; you will render an important service to natural theology, and shore up the basic principles that I tried to establish in my last two works.[15] (ibid., p. 14)

Receiving no reply, Bonnet wrote again on April 1, 1766 (ibid., p. 16), and an embarrassed Spallanzani responded on April 17 (Biagi, p. 84), telling him that he had already subjected more than two hundred aquatic salamanders to amputations, and a few thousand earthworms as well. "My main objective is to see if the new part [i.e., the regenerate] is a simple elongation of the old, or if it derives its origin from a tiny germ that is already present." There was then a lapse of several months in their correspondence, which caused Spallanzani some anxiety. When Bonnet finally responded, on August 8, 1766 (Castellani, p. 20), addressing several of Spallanzani's concerns, a floodgate was opened, and Spallanzani's next letter (September 21, 1766) contained a volume's worth of research observations and discussion.[16]

After writing a couple of pages detailing earthworm anatomy from his own studies, he abstracted for Bonnet a series of regeneration experiments on these "insects." For each experiment, a sentence gave the precise procedure, and the results were presented without comment. Spallanzani was also explicit about the need to replicate experimental observations and clearly understood the importance of approaching his studies without prejudice.

He continued with a description of an aquatic boat worm, as he called it, which was a bit larger than his earthworm. Again, he did a thorough dissection and described the vasculature in detail. He then

[14] "Nous n'admettrons pas que quelques molécules fortuitement réunies, produisent une queue, ou toute autre partie. Cette manière de philosopher serait le tombeau de la philosophie."

[15] "Une fausse philosophie s'éfforce de nous donner ces animalcules commes des bâtards de la Nature . . . multipliez tant que vous le pourrez les preuves de leur légitimation; vous rendrez un service important à la théologie naturelle, et nous étayerez les grands principes que j'ai tâché d'établir dans mes deux derniers ouvrages."

[16] The letter itself, in the archives of the Bibliothèque Publique et Universitaire in Geneva, is twenty-three and one-half pages long. Spallanzani included with it a separate "Programme" in which his earthworm regeneration studies were outlined.

performed regeneration experiments on them, using groups of fifty, and included sketches of their regenerates that unmistakably indicate the development and growth of regeneration blastemas.

Turning his attention to land snails, he subjected them to a variety of regeneration experiments. This led to his famous discovery of snail head regeneration, which gained much public notoriety. Voltaire was a member of the circle of correspondents to which Spallanzani had become attached through his relationship with Bonnet. Learning that Spallanzani had witnessed this remarkable phenomenon in land snails, Voltaire himself undertook to verify the observation. One of the results of his study was a satirical brochure entitled "Les colimaçons du révérend père l'Escarbotier, par la grace de Dieu Capucin indigne, . . . etc.," which Bonnet brought to Spallanzani's attention (Castellani, p. 86). Although this brochure helped to popularize the idea of head regeneration (see Goss's comments in Chapter 2 of the present volume), Spallanzani's observation was not universally accepted. He drew much sharp criticism for nearly ten years after his first report of snail head regeneration, especially from French natural historians.[17] Bonnet repeatedly came to his defense and eventually repeated and extended the snail studies, publishing his results in 1777.

To determine whether the reproductions were simple outgrowths of the old tissues or the expansion of preexisting germs, Spallanzani closely examined the interface between the stump and the regenerate. He found himself in a quandary, because, although he could not resolve this issue from his own observations, he did not believe that simple outgrowth could reasonably account for the regeneration of complex and organized appendages. He thus declared that it was "absolutely necessary to resort to the development of preexistent germs." Later, however, reflecting on his snail studies, he asked, among other questions, "Is it always necessary to resort to germs in order to explain their different reproductions?" (Biagi, p. 115). He was clearly more open-minded about the possibilities than Bonnet. Nevertheless, contemporary options left him no "rational" alternative to variations on Bonnet's preexisting germs.

Spallanzani then turned to regenerating tadpole and salamander tails, as another means of resolving the problem. He again provided Bonnet with a detailed description of the regenerative sequences and included several sketches to clarify his comments. (See Figures 1–8.) The conditions under which the animals were main-

[17] For a discussion of the philosophical differences that separated the French and Geneva schools of natural history in the eighteenth century, especially as it relates to Réaumur and Bonnet, see Dawson, 1987, pp. 25–48.

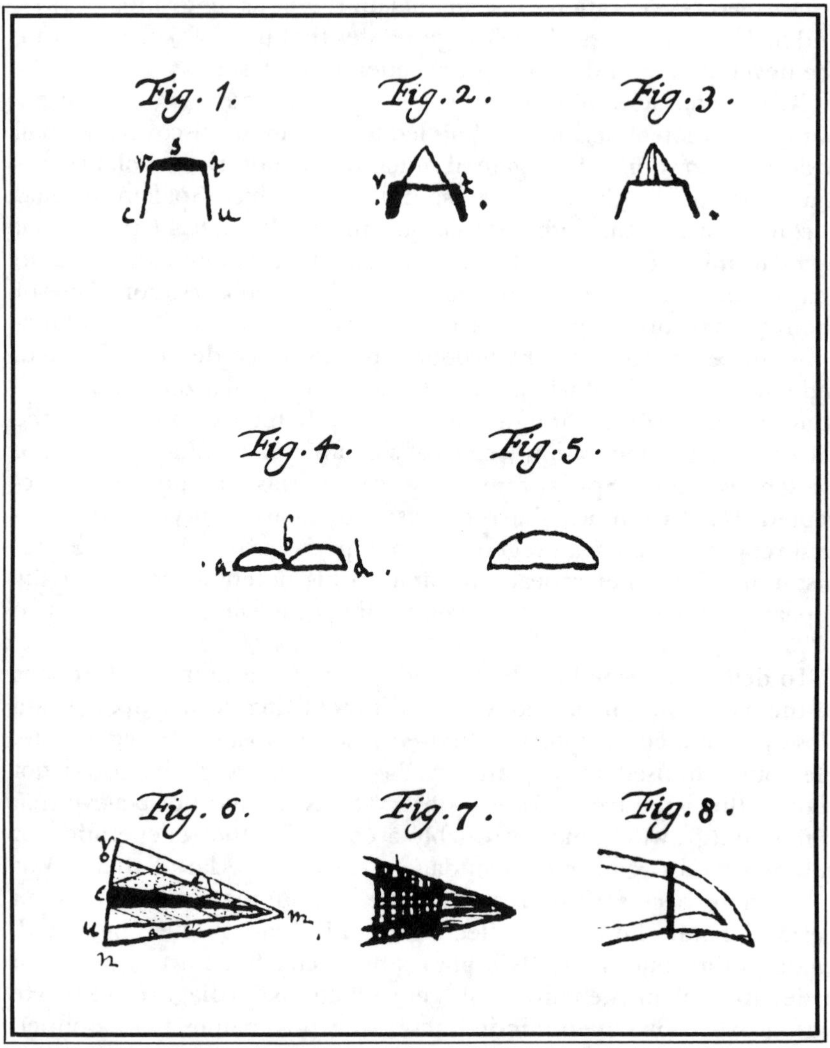

The first sketches of amphibian (both salamander and tadpole) tail regeneration, taken from Spallanzani's September 21, 1766, letter to Bonnet.

Fig. 1: The initiation of tail regeneration in an aquatic salamander. The stump is demarcated by *crtu* with the tapered regenerate forming at *rst*.

Fig. 2: Over time, the tail regenerate depicted in Fig. 1 elongates, terminating in a point.

Fig. 3: From the center of the stump, a brown band then appears extending to the tip, while the rest of the regenerate remains whitish. (We now recognize this as the regenerating caudal spinal cord.)

Fig. 4: In salamanders surviving complete tail amputation, regeneration is delayed. When it begins, two fleshy mounds appear, one on either side of the "spine" (located at *b*). (Caption continues on next page.)

tained, the time of year, and the failures as well as successes were diligently recorded.

An important comment concluded Spallanzani's observations on tadpole tails in this letter: "Nevertheless, I am almost led to believe that the tail regenerates in tadpoles are more of an elongation of the old parts than a development from a germ" (Biagi, vol. 1, p. 121). This comment was unlikely to have pleased Bonnet, and Spallanzani may have wrestled with the question of how to retain his friend's goodwill while preserving his own intellectual integrity. In the long run, however, Spallanzani did not abandon his ovist, preformationist view of the natural world. Although he had greater difficulty fitting regeneration into the preformation concept than did Bonnet, other "reasonable" explanations for regenerative phenomena were not yet available.

Over the next two years, they shared in their letters many thoughts on the nature of generation and regeneration. Spallanzani kept Bonnet immediately apprised of observations made in his vast array of experiments on animal regeneration, and in return Bonnet provided Spallanzani with a continuous flow of critiques and advice. An example of the latter will serve to bring this essay full circle to Trembley's pioneering work. In his advice to Spallanzani while the latter was developing the format for the *Prodromo,* Bonnet agreed that the work should be written in Italian, both for the sake of clarity and to keep peace with Spallanzani's colleagues at the university. But in recommending specific stylistic examples, he was quite clear: "The writings of Réaumur, Trembley, and Geer provide excellent models: by choice, imitate the second" (Castellani, p. 57).[18]

Caption to illustration *(cont.)*

Fig. 5: The mounds subsequently unite, and growth at point *b* removes the earlier inequality to form a single elevation.

Fig. 6: This schematic of a tadpole tail shows the fleshy core (*omu*) bounded by the gelatinous dorsal and fins (*vom, umn*). Spallanzani used it in explaining his observations on the circulation of blood in the tadpole tail.

Fig. 7: Contrary to expectation, the vasculature of the regenerated tadpole tail differed from that of the original: It formed a vascular network, instead of the single central artery and single central vein beneath the spinal cord in unamputated tails.

Fig. 8: Sometimes an abnormal tadpole tail regenerate was produced. These observations on tadpole tails appear to have led Spallanzani to doubt that Bonnet's "germes" could account for tail regeneration.

My thanks to Mr. Philippe Monnier, Adjunct Director of the Bibliothèque Publique et Universitaire in Geneva, for providing me with photocopies of the sketches from the Spallanzani–Bonnet correspondence.

[18] "Les Réaumur, Les Trembley, Les Geer, vous fournissent d'excellents modèles: imitez le second par préférence."

The work Spallanzani produced, however, was a far cry from the crisp, highly documented, and elaborately illustrated *Mémoires* Trembley had published. Although he might have created a comparable work, it appears that Spallanzani had other objectives in mind at that point in his career. Nevertheless, it may be of interest to reveal the data that lay behind the abridged product.

The Prodromo, *annotated*

Publication of the *Prodromo* established priority for Spallanzani in a wide range of discoveries on animal regeneration. The book was, however, simply a collection of brief essays. The promise of full disclosure of details in a later volume was never fulfilled. As already noted, observations and elaborate descriptions of protocols not included in the *Prodromo* can be found in the correspondence. The following paragraphs supplement his comments on earthworm and salamander tail regeneration in the *Prodromo* (from Maty's 1769 English translation) with the details he gave in his early letters to Bonnet. We can imagine what he might have said if he had been submitting his results to a modern journal.

Earthworms. A chapter in the *Prodromo* entitled "The Reproductions of the Earthworm" expanded upon studies Bonnet had begun in the early 1740s and discussed in his *Traité d'insectologie* (Paris, 1745), a work with which Spallanzani was familiar.

In the *Prodromo* Spallanzani began with a statement of method: He cut the worms transversely into three pieces. In the second sentence, he stated without further comment that the anterior part produced a tail and wondered if sectioning at different distances along the body would have any effect on the regenerative response. He then wrote, again without elaboration, "that nature has limits, which shall be determined in my work, and beyond which this reproduction of the tail can no longer be effected" (Spallanzani, 1769, p. 7). We are left to guess at the basis for these conclusions unless we turn to his correspondence.

In 1766, the summer heat favored his regeneration work, and in July he wrote to Bonnet that he would soon send an account of his discoveries. Spallanzani fulfilled that promise on September 21, 1766, detailing a vast array of regeneration experiments and observations and making rough sketches of structures in the process of regeneration. The letter was accompanied by a formal outline of a "Programme" around which he intended to organize for publication his data on earthworm regeneration. Here are some of the details he shared with Bonnet in that letter but omitted from the *Prodromo*.

He found the worms that he used in dung heaps. He demonstrated experimentally that they needed fresh air to survive, and he showed a rather modern understanding of scientific method by repeating the experiment to test the validity of his initial findings. Turning to blood vessels, he confirmed Bonnet's studies on the circulation of blood in worms and discussed the anatomy of their vasculature, having dissected more than a hundred specimens himself. His method of "anesthetizing" the animals to get a better view of their circulating blood was to keep them under water for about a day.

Having thus satisfied his own curiosity about the fundamentals of earthworm anatomy and physiology, even though he was familiar with the writings of others on these topics, he proceeded with his regeneration program. In his own words:

Here is the project that I set for myself, and that I have even finished. 1.) to cut a worm transversely into three parts such that the ovaries and little sacs [hearts?] remain in the anterior part, or that part where the head of the worm is located. *Result:* At the end of a few months, the posterior parts, or the tails, and the intermediate parts all died, except for 5 or 6 about which I will say more later. The anterior part which contained the ovaries and sacs, after 16–20 days began to develop a very little button [outgrowth], which stopped growing during the winter but which in the spring continued to develop, and at the present time the length of this regenerate in several worms is more than "deux pouces." [*Pouce* is a prerevolutionary unit of length approximately equivalent to 2.75 cm., or about 1 inch. (Dawson, 1987, p. 191, footnote)] 2.) Section in three parts with the ovaries and sac in the intermediate part. *Result:* All of the anterior parts or the parts with the head died. Several of the intermediate parts put forth a head on the anterior end and a tail on the posterior. In a few intermediate segments, I saw also that the head was the first to develop but it remains to be seen whether or not this observation is constant. In the Spring, the tail of these intermediate segments became very long but the head remained very short. These intermediate segments began to eat proving that the head was completely restored. In contrast, the tails died: except three which I put with the first experiment. (Biagi, vol. 1, p. 106)

Only after describing several more experiments on the earthworms, giving the results in the same format as in the passage just quoted and providing additional details as they were needed, did Spallanzani add important information that would have been stated at the beginning today: He had begun his experiments in September of the preceding year and had used two hundred animals for each experiment. Nevertheless, he concluded his discussion of his earthworm studies: "To argue from a stronger foundation and to increase further my own understanding of his obscure matter, I must repeat my proofs, vary them and invent new ones" (p. 107). He clearly understood the nature of good science.

Salamander tails. On November 9, 1765, Spallanzani announced his discovery of salamander tail regeneration to Bonnet and said that he would return to this fascinating subject for more detailed study in the spring. The elaborate letter written on September 21 of the following year contains significant details of his observations. We can again begin with the *Prodromo* to see what his published observations were.

The chapter entitled "Reproductions of the Tail in the Aquatic Salamander" opens with nearly seven pages of observations on the natural history of salamanders, their need for air, their resistance (or lack thereof) to heat and cold, and other issues. Spallanzani then turns to his tail-amputation studies, dividing his observations on regenerative phenomena into seven classes or sets of questions, beginning as follows: "I. Does the reproduction take place, 1. in every known species of salamander? 2. And at any period of their life? 3. Does it happen in whatever situation they may be kept, upon the earth, or in water? 4. Is it brought about, let the length of the divided part be greater or less?" (Spallanzani, pp. 64–5). To these questions he answers with an unequivocal but unexplained yes, and he moves immediately to the next set of questions. This Spartan mode of disclosure is reiterated six more times as he essentially lists findings on the following subjects: differences between the amount cut off and the amount regenerated; age- and species-dependent differences in regenerative ability; anatomy of the regenerate relative to the original; the proportion of vertebrae regenerated; the time course of tail regeneration; seasonal variation in the regenerative rate; differences in regenerates, engendered by different axes of amputation; the effects of sectioning only the spinal cord or the musculature; and repeated amputation. A sentence closes this last topic and the chapter: "The same process takes place not only in a second reproduction, but also in a third, in a fourth, etc. and the salamanders deprived of many successive reproductions, still follow in the formation of new parts, the same unalterable laws" (p. 67). But what is the basis for this surfeit of conclusions?

We again turn to Spallanzani's prepublication correspondence with Bonnet, especially the September 21, 1766, letter, and here is a summary of what he told Bonnet. Spallanzani collected his salamanders from ditches and pits. In the spring (he soon found that the regenerative rate was seasonably variable), he amputated the tails of full-grown salamanders, and then, at midsummer, he turned to month-old larvae. This is probably the basis for his declarations in the *Prodromo* on age-dependent variability in regeneration. Profuse bleeding attended the amputation of tails in the adults. He described superficial wound healing of the stump quite accurately and continued his report to Bonnet with a temporal sequence of observations on the formation and growth of the tail regenerate. This was accompanied by labeled sketches of the interface between tail stump and regenerate and a commentary on the progressive darkening of skin color on the regenerate until it was nearly as dark as that of the original tail. At first, he did not know what to feed the salamanders, so he fed them nothing and observed that this did not have an obvious effect on their regeneration. Concerned with the question of whether the regenerate arose from a "prolongation of the fibers" of stump tissues or from preformed "germes," he dissected regenerates. He had not resolved the issue by the time he wrote in September, but by November 23, 1766, when he next wrote to Bonnet, he had the answer: "I can now tell you that in all of the tail regenerates I have found the bones of which the vertebral column is made. They are inserted into each other just like in the original tail. There is no difference between the 'artificial' vertebrae and the native ones" (p. 127).[19]

In conclusion, this essay has drawn attention ultimately to the first five years of the epistolary relationship between Spallanzani and Bonnet. The contents of their letters in that period bring into relief the importance of regeneration research both to Spallanzani's intellectual development and to transitions in eighteenth-century natural history and philosophy. The former is revealed by the number of organisms Spallanzani employed in a wonderfully diverse array of experiments aimed at resolving issues of "reproductions." His enthusiastic comments to Bonnet indicate how completely engaged he was in pursuit of understanding animal regeneration.

As to the latter, detailed discussions of methodology in their letters reveals a clear emphasis on the new experimental paradigm. Their promotion of empirical methods as the basis of a valid natural philosophy, coupled with the simplicity of fundamental regeneration experiments (e.g., snail head amputation), promoted a spreading wave of activity in this domain.

[19] "A présent je vous signifie, que dans toute les reproductions de ces queues j'ai trouvé les osselets, dont cette épine est composée. Ils sont enchassés les uns dans les autres, comme dans la queue naturelle. Point de différence entre ces osselets artificiels, et les naturelles."

The foregoing analysis abstracts only a few of Spallanzani's observations and comments related to his regeneration research. Although they were made more than two hundred years ago, they form a cornerstone upon which contemporary regeneration research is built. Thus, the flowering of regeneration research in the eighteenth century, prompted by Lazzaro Spallanzani's example, marks an important period in the evolution of modern experimental zoology.

REFERENCES

Biagi, Benedetto, ed. 1958. *Lazzaro Spallanzani: Epistolario.* 5 vols. Sansoni Antiquariato, Florence.
Bonnet, Charles. 1745. *Traité d'insectologie.* Duvand Librairie, Paris.
   1777. Expériences sur la régénération de la Tête du Limaçon terrestre. *Journal de Physique,* September 1777, 10(2): 165–79.
Castellani, Carlo, ed. 1971. *Lettres à M. l'abbé Spallanzani de Charles Bonnet.* Episteme Editrice, Milan.
Dawson, Virginia P. 1984. *Trembley, Bonnet, and Réaumur and the Issue of Biological Continuity. Studies in Eighteenth-Century Culture* 13: 43–63.
   1987. *Nature's Enigma: The Problem of the Polyp in the Letters of Bonnet, Trembley, and Réaumur.* American Philosophical Society, Philadelphia.
Dolman, Claude E. 1975. Spallanzani, Lazzaro. *Dictionary of Scientific Biography,* 12: 553–67.
Hankins, Thomas L. 1985. *Science and the Enlightenment.* Cambridge University Press, Cambridge.
Lenhoff, Sylvia, and Howard Lenhoff. 1986. *Hydra and the Birth of Experimental Biology–1744: Abraham Trembley's Memoirs Concerning the Natural History of the Freshwater Polyp with Arms Shaped Like Horns.* Boxwood Press, Pacific Grove, Calif.
Montalenti, Guiseppe. 1976. Vallisnieri, Antonio. *Dictionary of Scientific Biography,* 13: 562–5.
Pilet, P. E. 1973. Bonnet, Charles. *Dictionary of Scientific Biography,* 2: 286–7.
Roe, Shirley A. 1981. *Matter, Life and Generation: Eighteenth-century Embryology and the Haller–Wolff Debate.* Cambridge University Press, Cambridge.
Roger, Jacques. 1963. *Les sciences de la vie dans la pensée française du XVIIIe siècle.* Armand Colin, Poitiers.
Rostand, Jean. 1951. *Les origines de la biologie expérimentale et l'abbé Spallanzani.* Fasquelle Editeurs, Paris.
Sandler, Iris. 1973. The reexamination of Spallanzani's interpretation of the role of the spermatic animalcules in fertilization. *J. Hist. Biol.* 6: 193–223.
Savioz, Raymond, ed. 1948. *Mémoires autobiographiques de Charles Bonnet de Genève.* J. Vrin, Paris.
Spallanzani, Lazzaro. 1765. *Dissertazioni due.* Bartolomeo Soliani, Modena.
   1768. *Prodromo di un opera da imprimersi sopra la riproduzioni animali.* Giovanni Montanari, Modena. Translated 1769 by M. Maty, *An Essay on Animal Reproduction.* T. Becket & DeHondt, London, 1769.

Trembley, Abraham. 1744. *Mémoires, pour servir à l'histoire d'un genre de polypes d'eau douce, à bras en forme de cornes.* Verbeek, Leiden. Lenhoff and Lenhoff (1986) contains a complete English translation.
Wiener, Philip P., ed. and trans. 1951. *Leibniz: Selections.* Scribners, New York.

6

Observation versus philosophical commitment in eighteenth-century ideas of regeneration and generation

KEITH R. BENSON

The standard depictions of eighteenth-century disputes concerning animal generation center on two rival positions: preformation and epigenesis.[1] The preformation doctrine, which frequently received remarkable corroboration from experimental demonstrations and observations of natural phenomena, became almost universally popular at the end of the seventeenth century. Much of the enthusiasm and support came from novel microscopical studies that suggested the existence of preformed organisms within embryonic tissue or, at the very least, preformed rudimentary structures within the embryo. The second view, epigenesis, championed in its Aristotelian guise by William Harvey (1578–1657) in the seventeenth century and then translated into the new mechanical framework by René Descartes (1596–1650), asserted that new organisms arose from previously undifferentiated matter as a completely new generation.[2] The stan-

[1] The standard references detailing the epigenesis–preformation debate include the following: Charles W. Bodemer, "Regeneration and the decline of preformation in eighteenth-century embryology," *Bulletin of the History of Medicine*, 38 (1964):20–31; Peter J. Bowler, "Preformation and pre-existence in the seventeenth century: A brief analysis," *Journal of the History of Biology*, 4 (1971):221–44; F. J. Cole, *Early Theories of Sexual Generation* (Oxford: Oxford University Press, 1930); Elizabeth Gasking, *Investigations into Generation, 1651–1821* (Baltimore: Johns Hopkins University Press, 1967); Shirley A. Roe, *Matter, Life, and Generation: Eighteenth-Century Embryology and the Haller–Wolff Debate* (Cambridge: Cambridge University Press, 1981); and Jacques Roger, *Les sciences de la vie dans la penseé française du XVIIIe siècle* (Paris, 1963). For an overview of these works, excepting Roe's, and a critical appraisal of the simplistic epigenesis–preformation debate, see Keith R. Benson, "John Turberville Needham (1713–1781) and Eighteenth-Century Theories of Generation," M.A. thesis, Oregon State University, 1973.

[2] William Harvey, *Exercitationes de generatione animalium* (London: O. Pulleyn, 1651), and René Descartes, *L'homme et un traité de la formation du foetus* (Paris: Charles Angot, 1664). Both works have been translated into English and are more accessible in these editions. See William Harvey, *Works*, trans. Robert Willis (London: Sydenham Society, 1847; facsimile reprint ed., New York: Johnson Reprint, 1965), and René Descartes, *Treatise on Man*, French-English edition, ed. Thomas Steele Hall (Cambridge, Mass.: Harvard University Press, 1972).

91

dard account often relates how the preformation view, with its encapsulated spermatic homunculus or its encased adult forms within the egg, became slowly eroded in the eighteenth century by observations and experiments of Enlightenment natural philosophers, particularly those who examined major problems in animal generation with carefully conceived experiments or painstaking microscopical observations. This led to a new position, usually described in historical treatments as a refined version of epigenesis, which represented another triumph of reason and demonstrated the success of the new scientific methods from the eighteenth century.

Undoubtedly, much of the credit for this simplified view of eighteenth-century theories of generation stems from the position of that century vis-à-vis the scientific revolution. It was, as Thomas Hankins asserts, the century in which natural philosophers completed the replacement of the Aristotelian world view with a completely mechanical philosophy in the tradition of Descartes and Newton.[3] The new tradition, more modern and understandable to the contemporary mind, has received the mantle of conqueror, gradually overcoming the entrenched but outmoded idea of preformation by providing examples from nature of how the natural world really works. That is, as was typical of the new approaches to science in the eighteenth century, phenomena associated with animal generation were described with care and accuracy, thus leading inevitably to the formulation of the basic outlines of epigenesis.

Unfortunately, for this is a compelling story, the *philosophe* at the end of the eighteenth century was just as confused concerning animal generation as was the *savant* of the seventeenth century. That is, despite many observations and experiments, confusion reigned supreme in interpreting the results of this work. Arnulphe d'Aumont, writing in the 1765 edition of Diderot's *Encyclopédie* on "Generation," provides some indication of the problematic status of animal generation: "It [generation] is, at present, a mystery for us into which we are so little advanced that the manifold attempts at explanations only serves to convince us more and more of their futility."[4] Across the English Channel, natural philosophers were equally mystified. The *Encyclopaedia Britannica*, the repository of the latest wisdom from the British mind, provided a summary of its discussion concerning generation, in 1771, with the following pessimistic note:

[3] Thomas L. Hankins, *Science and the Enlightenment* (Cambridge: Cambridge University Press, 1985).

[4] Arnulphe d'Aumont, "Generatione," *Encyclopédie ou Dictionnaire raisonné des sciences*, 1765 edition, vol. 17.

Whoever reads this short sketch of the different theories of generation that have hitherto been invented, will probably require no other arguments to convince him, that physicians and philosophers are still as ignorant of the nature of this mysterious operation as they were in the days of Noah.[5]

In fact, there was no simple preformation versus epigenesis skirmish during which one position vanquished the other on a field of battle. Instead, throughout the eighteenth century ideas of generation fluctuated from preformation of the animalculist version, to variants of ovism, to attempts at reinterpreting Harverian epigenesis within a novel Newtonian mechanical framework. What is particularly interesting about these debates over generation is the comparative role of observation versus philosophical argument in determining the character of the controversy. And nowhere was there more excitement and confusion than in discussions surrounding observations of the regeneration of lost parts of animals or the regeneration of whole animals. Like hybrids and monsters, two other anomalies of animal generation that exercised the minds and pens of the *philosophe*, regenerated structures and organisms offered a potential demonstration from nature of how generation proceeded.[6] At the very least, it provided a difficult problem with which the variety of rival positions had to contend.

Regeneration became a major issue after the popularization of Abraham Trembley's (1710–84) observations on the phenomenon he observed in freshwater hydra, published in his *Mémoires* in 1744. Actually, Trembley communicated his hydra work to colleagues through numerous letters in 1740 and was generously cited by them until his own work appeared four years later.[7] Additionally, there were a number of earlier reports of regeneration, some of which clearly anticipated the problems it was to create for ideas of animal generation. Melchisedech Thevenot (1620 or 1621–92) and Claude Perrault (1613–88), two of the original members of the French Academy of Science, both observed tail regeneration in lizards during a series of observations completed under the auspices of the academy in the 1680s.[8] Neither wrote extensively on these observations. In the case of Perrault, however, these observations may have led him to move from the simple and then-popular concept of *emboîtement*, or the encase-

---

[5] "Generation," *Encyclopaedia Britannica*, 1771 edition, vol. 2.

[6] Roe, *Matter, Life, and Generation*, pp. 1–20.

[7] I thank Charles E. Dinsmore for the details concerning Trembley's communication with his colleagues.

[8] The most complete sketch of Thevenot is C. Stewart Gillmor, "Thevenot, Melchisedech," in *Dictionary of Scientific Biography* (hereafter *DSB*), 13: 334–7. For a sketch of Perrault, see A. G. Keller, "Perrault, Claude," *DSB*, 10: 519–21.

ment of completely formed organisms within the germ, which Jan Swammerdam (1637–80) had popularized through his butterfly observations, to the speculation that embryonic growth resulted from preformed germs that were present in all parts of the body. Perrault had recognized that regeneration of parts could not be explained by the simple version of encasement, because this view included only the encapsulation of the whole organism. To maintain a logically consistent version of preformation that remained within the mechanical philosophy of the seventeenth century required that each regenerating structure in the organism contain a miniaturized form of itself or some "germ" of itself. Perrault's speculative adventures stemmed from his acceptance of the Leibnizian notion that all parts of the organic machine possess vital properties. The preformed germs were such entities, thereby providing Perrault with organic particles to explain regenerated structures.[9]

Another early adherent of preformation was René-Antoine Ferchault de Réaumur (1683–1757). Typical of many eighteenth-century natural philosophers, he accepted the mechanical philosophy and preformation and rejected Aristotelian epigenesis with its antiquated notions of causation and qualitative change. But Réaumur was also a first-rate observer, and as a result his version of preformation was neither simpleminded nor without an empirical base.[10] As early as 1712, he noted that biparental inheritance required a more sophisticated version of preformation than *emboîtement*. In this same year he published a paper in which he included his observations on the regeneration of claws in crustaceans.[11] Here, again, the phenomenon pointed to the inadequacy of the simple view of encasement. Noting the same problems Perrault recognized with *emboîtement*, Réaumur nevertheless continued to adhere to a modified version of preformation, because the epigenetic alternatives were vexed with greater problems. How could a precise and differentiated structure like a leg regenerate spontaneously from undifferentiated material without some regulative guidance? For Réaumur, the only acceptable explanation that was consistent with the mechanical philosophy was a variant of preformation in which the body of the crayfish contained invisible "germs" scattered throughout it. Amputation of any part provided the stimulus for the growth of the germs into the proper replacement

---

[9] Roger, *Les sciences de la vie,* pp. 346 and 368.
[10] See Chapter 3 in the present volume.
[11] Roe, *Matter, Life, and Generation,* p. 9, and J. B. Gough, "Réaumur, René-Antoine Ferchault de," *DSB,* 11: 327–35.

part. These preformed germs were present at the original Creation of the crayfish and, therefore, were of limited quantity; parts could not regenerate infinitely.

Nicolaas Hartsoeker (1656–1725), the Dutch instrument maker and self-trained naturalist, also investigated regeneration after presenting strong arguments in favor of spermatic preformation in his *Essai de dioptrique* (1694). But thirty years later, Hartsoeker changed his position when he published *Recueil de plusieurs pièces* (1722), containing his own work on crustacean regeneration.[12] For him, a regenerated limb was a new creation, and both his observations and Réaumur's discoveries demonstrated unequivocally that nature provided new creations via regeneration daily. Therefore, critics who attacked epigenesis on the grounds that it required a new creation for each generative act were arguing a moot point. Generation, as demonstrated through the regeneration of lost parts, was a new creation. As Hartsoeker concluded, "The intelligence which can reproduce the lost claw of a crayfish can reproduce the entire animal."[13]

By the mid-1720s, therefore, several cases of regeneration had been noted and reported. But because of the recognized problems that regeneration raised for both preformation and epigenesis, there was no universal agreement on an explanation. The event that drew new attention near midcentury to the problem was Abraham Trembley's observations of regeneration in hydra.[14] Working on the estate of his benefactor, Count Bentinck, Trembley initially observed regeneration in a series of experiments on the freshwater hydra.[15] Trembley cut the hydra in half to determine if the plant-like organism was indeed a plant; enigmatically, it possessed tentacles that behaved in an animal-like fashion. The result of the experiment was that both parts regenerated into two indistinguishable hydra. Trembley's taxonomic hunch, as a result, was that the hydra was an animal. Lacking complete confidence in his conclusions, he wrote to Réaumur to seek his advice. Responding from Paris, Réaumur concurred with Trembley's assumption that the hydra was indeed an animal. Moreover, he arranged to have Trembley's paper read before the Académie Royale des Sciences in 1741. Both the letter and subsequent demonstrations of the hydra's unusual behavior caused a notable stir in Paris. Eventually Trembley published *Mémoires, pour servir à l'historie d'un genre de polypes d'eau*

12 Benson, "Needham," p. 24.
13 Hartsoeker, quoted in Gasking, *Investigations into Generation*, p. 86.
14 Roe, *Matter, Life, and Generation*, pp. 10–11.
15 See Chapter 4 in the present volume.

douce, à bras en forme de cornes (1744), the same year the academy issued its own report. The latter indicates the excitement over Trembley's discovery:

The story of the Phoenix that is reborn from its ashes, wholly fabulous as it is, offers nothing more marvellous than the discovery of which we are about to speak. The chimerical ideas of palingenesis or regeneration of plants and animals, which some alchemists have thought possible by the assembly and re-union of their essential parts, only tended to restore a plant or an animal after its destruction; the serpent cut in two and said to join together again, only gave one and the same serpent; but here is nature going farther than our fancies.[16]

The distance nature had gone was not only impressive to Trembley's contemporaries; it was also perplexing. And when Charles Bonnet (1720–93) published *Traité d'insectologie* (1745), with similar observations on segmented worms, regeneration became a cause célèbre that demanded an explanation from the various eighteenth-century views of generation.[17] As the historian Aram Vartanian stated, "In the pieces of a cut-up polyp regenerating into new polyps, Trembley's contemporaries had the startling spectacle of Nature caught, as it were, *in flagrante* with the creation of life out of its own substance without prior design."[18] Immediately, the problem of regeneration served as grist for the mills of the *philosophes*, and soon the observational work was swamped with philosophical discussion.

If the hydra demonstrated the creation of designed life from its own substance that lacked design, where was the regeneration? More critically for the philosophers, if each part of an animal could regenerate the entire animal, where was the residence of the "soul," the recognized organizing principle of organic beings? Julien La Mettrie (1709–51) and Denis Diderot (1713–84), also borrowing heavily from the German philosopher Leibniz, used the hydra to argue against the notion of the soul as an organizing principle or as the universal criterion for life.[19] Instead, they believed regeneration illustrated that life and organization were properties resident throughout organic matter. Neither La Mettrie nor Diderot was a naturalist, and their position quickly drew fire from naturalists concerned with the materialistic overtones of their views.

---

[16] Quoted from *Histoire de l'Academie Royale des Sciences, Année 1741* (1744), pp. 33–4 in Roe, *Matter, Life, and Generation*, p. 10.
[17] Roe, *Matter, Life, and Generation*, p. 23.
[18] Aram Vartanian, review of *Abraham Trembley of Geneva: Scientist and Philosopher*, by John R. Baker, *Isis* 44 (1953): 388.
[19] Hankins, *Science and the Enlightenment*, p. 133.

Georges-Louis Leclerc, comte de Buffon (1707–88), the politically powerful and literary nobleman in charge of the Jardin du Roi in Paris, entered the fray backed by his experimental work, conducted primarily by the English microscopist and cleric John Turberville Needham (1713–61).[20] Buffon, anxious to demonstrate that organic molecules were preexistent to life and were subsequently acted upon by his conception of the force-like *moule intérieur* (internal mold) to organize the material into the various forms of life, was an adherent of epigenesis. Escaping the materialist position of La Mettrie and Diderot, he accounted for design in nature through the specificity of the *moule intérieur*. This idea was ready-made for epigenesis, since a lost part could be regenerated because the *moule intérieur* had restored the integrity of the organism by reorganizing organic molecules resident within the organisms.

But if Buffon could provide ad hoc methods to "explain away" problems for epigenesis, the preformationists could do the same. Albrecht von Haller (1708–77), the Swiss physician and polymath, was also an early adherent of epigenesis. Arguing against several preformationists who used microscopical observations to demonstrate their position, Haller quoted from Aristotle and Harvey to support his view that generation occurred epigenetically from undifferentiated organic fluids. In *Primae lineae physiologiae* (1747), he adopted a Buffon-like force explanation, arguing by analogy that the process was under the control of Newtonian laws of attraction. Generation was regulated, he claimed, "undoubtedly by divine laws, which in their proper manner order spicules of ice, crystals of salts, particles and sheets of metals . . . to be joined together."[21]

But Haller soon became skeptical about epigenesis, partly because of his investigations into the embryological development of the chick. In *Elementa physiologiae*, written some ten years after his profession of epigenesis, Haller changed positions. From his examination of the early development of the chick embryo, he could observe no stage in which the maternal tissue was not continuous with the embryonic tissue, indicating that a part of the embryo was preexistent to the formation of the actual organism. He also relied on the regeneration studies of his countryman, Charles Bonnet, when he adopted an ovist position of preformation in which the egg contained preformed parts and the sperm provided the activation for the coalescence of the parts.

---

[20] Benson, "Needham," and Roe, *Matter, Life, and Generation*, both detail the relationship between Buffon and Needham and describe the observational work resulting from their collaboration.

[21] Quoted in Roe, *Matter, Life, and Generation*, p. 24.

Clearly, however, Haller was persuaded primarily by philosophical considerations. Observing that it might appear as if the parts of an organism emerged gradually from undefined material, he argued that this empirical sighting was not reliable because epigenesis was impossible: "As neither volition of the individual, nor chance, nor a blind force imparting movement to the organic parts is able to form the organism, we have no choice but to admit that the embryo is already formed before fecundation."[22] In Haller's case, where observation was ambiguous, philosophical argument was to be relied upon. And the only rational view within the mechanical tradition for Haller was preformation.

Haller owed much to Charles Bonnet (1720–93) for his conversion to preformation. It was Bonnet, the quintessential empiricist cum philosopher of the eighteenth century, who dealt with regeneration. Bonnet began his interest in the subject after reading the reports of Trembley (his cousin and trusted friend) and Réaumur, and extended their studies to segmented worms. Observing for himself that regeneration was a widespread phenomenon of nature, he adopted a refined position of preformation to replace the simple *emboîtement* concept. His position included an expanded notion of the word "germ":

The term *emboîtement* suggests an idea which is not altogether correct. The germs are not enclosed like boxes within the other, but a germ forms part of another germ as a seed is a part of the plant on which it develops.... I understand by the word "germ" every preordination, every preformation of parts capable by itself of determining the existence of a plant or animal.[23]

Equipped with this philosophical position – for the germs were not visible – Bonnet had no problem with regeneration, and in his mind the hydra became the archetypal example of an organism regenerated by the germ that was defined as "any secret preorganization."

Bonnet's scientific career was cut short by deafness and blindness, but these afflictions did not prevent him from shifting to even more philosophical and metaphysical concerns by midcentury. His ruminations pointed to the philosophical flaws of epigenesis: It implied spontaneous generation; it indicated that nature allowed new generation; and it had a materialistic overtone, suggesting that inorganic substances could form organic beings. Growing increasingly confident of his position and contemptuous of epigenesis, he used the hydra to attack those who still held to epigenesis, especially Buffon and Voltaire:

[22] Quoted in Gasking, *Investigations into Generation*, p. 91.
[23] Quoted in Cole, *Early Theories*, p. 99.

We do not wish to have recourse to purely mechanical explanations, which experience does not justify and which good philosophy condemns, we must think that the polyp is, so to speak, formed by the repetition of an infinity of small polyps, which only await favorable conditions to come forth.[24]

Enlisting the aid of his new friend, Lazzaro Spallanzani (1729–99), Bonnet encouraged him to extend his regeneration studies to the earthworm, which Bonnet had studied earlier. Spallanzani, like Bonnet, had been arguing with Buffon and Needham over the notions of epigenesis and spontaneous generation. Also similar to Bonnet, Spallanzani was a great believer in observation, especially using the microscope. Beginning the research in 1765, he eventually extended the studies to include slugs and frog tadpoles, noting especially the changes in the regenerative site following the excision of a particular part. In 1766 he also communicated to Bonnet his discovery of snail and salamander regeneration, work that Bonnet considered novel.[25] These observations were published in Spallanzani's influential *Prodromo di un opera da imprimersi sopra la riproduzioni animali* (1768). Combining the regeneration work with his observations on limb development and embryonic development in amphibians, Spallanzani clearly argued for preformation. Significantly, he depended on philosophical judgment and logical argument in addition to observation:

I should reply without hesitation, . . . all Nature abounds with such evolutions, according to the accounts of the most judicious philosophers of this age, it is natural to suppose, that these orders of fetuses, which annually make their appearance in the ovaria, are not successively generated, but co-existed with the female, and are only unfolded, and rendered visible in progress of time, by the supplies of nutritive liquor that come from the female. This coexistence of successive orders of fetuses, which become visible in the ovaria, is analogous to that which takes place in the limbs. . . . Is it not infinitely more philosophical to suppose, that the limbs coexist with the tadpoles, and are invisible, only because they are too small to strike the senses? And if it is reasonable, to adopt this opinion concerning the limbs, shall we not also admit it with respect to the fetuses of these animals?[26]

Bonnet read the results of his colleague's work, first praising him for his unbiased observations: "You have cherished no theory, but are

---

24 Quoted in C. O. Whitman, "The palingenesia and the germ doctrine of Bonnet," *Wood's Hole Biological Lectures, 1894*, p. 12.

25 I thank Charles E. Dinsmore for this information on the Bonnet–Spallanzani correspondence.

26 Spallanzani, abbé [Lazzaro]. *Dissertations Relative to the Natural History of Animals and Vegetables*, 2 vols. (translated by T. Beddoes from the Italian.) (London: John Murray, 1784, 1789), 2: 89–90.

satisfied with interrogating nature, and giving the public a faithful account of her responses."[27] Bonnet then incorporated the work into his own prose and presented his own version of an argument for preformation:

Here there is a new organized whole, which grows from an ancient one, and constitutes the same body; there is an animal slip that grows, and expands itself on the stump of an animal as a vegetable slip does on the trunk of a tree. Remark that the flesh of the piece cut off does not in the least contribute to the formation of the part regenerated; the stump only nourished the bud; it being the soil in which the latter vegetates. The part then that is reproduced passes through all the degrees of growth, by which the entire animal itself had passed. It is a real animal in a very minute form in the great animal that served for a matrix.[28]

Clearly, Spallanzani and Bonnet were in the company of all the natural philosophers of the eighteenth century. Typical of the period, reason and experience were inextricably bound – and where one was lacking the other often prevailed. Neither natural philosopher ever observed the germs he so confidently used to explain regeneration, but for both of them the problems associated with epigenesis were insuperable. These same problems they considered to be easily explained by the hypothetical germs associated with the preformationist position.

The problem of animal generation presented the *philosophe* with the "mystery of mysteries." Preferring mechanical explanations and eschewing unknown and unobservable forces, many *philosophes* refused to abandon the mechanical version of preformation, because epigenesis was too closely aligned with the outmoded Aristotelian world view. Others, however, especially those who adopted the work of Harvey but replaced the Aristotelian explanations with Newtonian lawlike forces, preferred epigenesis. Throughout this period, both positions offered viable frameworks for understanding generation. But as the studies of animal regeneration suggest, the crucial test of the theory was not observation. Indeed, it is a grave mistake to overemphasize the role of observation among the *philosophes* who dealt with regeneration. Beginning with empirical investigations, all the natural philosophers who studied regeneration ultimately accepted preformation or epigenesis on the basis of philosophical considerations.

---

[27] Quoted in Claude E. Dolman, "Spallanzani, Lazzaro," *DSB,* 12: 553–67.
[28] Quoted from Charles Bonnet, *Contemplation de la nature* (Amsterdam, 1769), in T. S. Hall, *A Source Book in Animal Biology* (New York: McGraw-Hill, 1951), p. 378.

7

The neurotrophic phenomenon: its history during limb regeneration in the newt

MARCUS SINGER and JACQUELINE GÉRAUDIE

T. J. Todd's discovery of the neurotrophic phenomenon

In 1823 an English physician, Tweedy John Todd (1789–1840), reported in the *Quarterly Journal of Science, Literature and Arts* the following experiment on the ability of the aquatic salamander (newt) to regrow an amputated limb:

> If the sciatic nerve be intersected at the time of amputation, that part of the stump below the section of the nerve mortifies. . . . If the division of the nerve be made after the healing of the stump, reproduction [regeneration] is either retarded or entirely prevented. And if the nerve be divided after reproduction has commenced, or considerably advanced, the new growth remains stationary, or it wastes, becomes shrivelled and shapeless, or entirely disappears. This derangement cannot in my opinion, be fairly attributed to the vascular derangement induced in the limb by the wound of the division, but must arise from something peculiar in the influence of the nerve. (pp. 91–2)

This clear statement of experimental fact was possibly the first proof in a scientifically controlled experiment that nerves, in addition to their functions of impulse conduction and transmission, may also serve to maintain the morphological and developmental integrity of the organs they serve. Such a view was expressed by Samuels (1860). (See review in Wyburn-Mason, 1950.) This activity was called "trophic" by Parker in 1932, from the Greek term meaning "nourishing." More recently the term "neurotrophic" has been used to specify the nervous source of the activity. (See reviews in Guth, 1969, and Drachman, 1974.) In recent years the term "neurotrophic" has been preempted to describe the effect of nerve growth factor (NGF) on growing nerve fibers. Although the older usage has been adequately criticized by Brockes (1987), we use the term, throughout this chapter, in its old-fashioned meaning.

The next substantive experiment on the neurotrophic phenomenon was reported in 1876 by Vintschgau and Honigschmied (see

101

Guth, 1971), who recorded the morphological dependence of the vertebrate taste bud on the sensory nerve supply. They were unaware of Todd's paper of some fifty years earlier. It was only in the early part of this century, about one hundred years after the publication of Todd's article, that his work was noticed. Indeed, T. H. Morgan, in his classic book on regeneration published in 1901, also was unaware of Todd's work and made no reference to the "neurotrophic" phenomenon. The eminent physiologist Johannes Müller (1801–58), in his *Elements of Physiology* (1843), did refer to these experiments; presumably Todd's, but Müller did not remember the source of his information or the name of the author. Müller's laboratory in Berlin was often visited by scientists; perhaps Todd had stopped there. Müller himself also traveled frequently to other centers abroad; perhaps on one of these visits Müller learned of Todd's work. He did cite the results of the unnamed researcher as experimental proof of the "nourishing" function of nerves.

Müller also referred to experiments showing that the spinal cord is necessary for regeneration of the newt's tail (again without mentioning the original source). Todd had specifically noted that cutting the spinal cord *at the base of the tail* did *not* prevent regeneration of the newt tail. In his studies of the neurons' influence on tail regeneration, Todd cut the spinal cord at the base of the tail but found that, unlike in his work with the limb, he did not observe suppression of regeneration. Apparently he assumed that all nerves in the tail were destroyed by severing the spinal cord. He did not know, as we do now, that only the descending and ascending fibers were interrupted and that many neurons are present in the tail portion of the cord. These were responsible for the new growth; for had he avulsed the cord to the end of the amputation surface and blocked regeneration of the cord by some impediment (such as gutta-purcha, which was later used), he would have aborted regeneration. The spinal cord in the salamander does regenerate new neurons, roots as well as long tracts (Singer, Nordlander, and Egar, 1979; Géraudie, Nordlander, and Singer, 1988). But even with this citation and Müller's recognition of the significance of his work, Todd's experiments on the nerve had no meaningful influence on trophic concepts until almost the middle of the present century. One reason is that discussions of the "nutritive" function of the nerve revolved around clinical examples, particularly the wasting of the paralyzed limb. Moreover, animal experiments were not usually designed to serve as models of human dysfunctions.

One wonders why Todd performed his experiment in the first place. In his article he gave no reason. The answer may lie in the way in which the structure and function of the nerve was then understood. In the early nineteenth century, microscopic anatomy was in its infancy. The tissues and organs of the body were generally considered a syncytium – the nerve, for example, blending and becoming continuous with the muscle. The cell theory of Schleiden and Schwann – that the cell is the physiological, morphological, and genetic unit of the body – had not yet been formulated. It was not until the 1830s that histological research came into favor and techniques of tissue preparation and staining were introduced as a routine research tool. Hence the fiber makeup of nerve was unknown to Todd. Indeed, the neuron doctrine – the idea that the neuronal cell body and its processes constitute a single cell – was not proposed until much later in the century. In his day the ancient view that nerves are tubular conduits transmitting a vital principle in the form of an essence, fluid, particle, or other substance of motion or sensibility from one region to another was still current. In the case of muscle it was often assumed that the humoral principle (*vis nervosa*) activates muscle and sustains (nourishes) the muscle substance. The English medical doctor must also have been aware of the then-popular view of the primacy of the nervous system in maintaining normal health and development. This theory had its origins in the seventeenth- and eighteenth-century writings of Willis (1659), Hoffmann (1761), Unzer (1771), and Thaer (1774) (reviewed in Rath, 1959). Elements of this notion persist today in more modern form as psychosomatic medicine. The theory reached its highest development in the discourses on a system of neuropathology (1776) by the eminent Edinburgh physician William Cullen (1740–90), who suggested that nervous disorders are the basis for all disease, even in non-nervous organs (Rath, 1959). It was only after Todd's time, in the middle of the nineteenth century, when microscopy revealed the cellular nature of tissues and organs in both health and disease, that the new system of pathology by Virchow replaced the neuropathology of Cullen and of other "neurists," as Virchow called them. (See Rath, 1959.) In Virchow's new pathology, the nervous system is only one possible starting point of disease; any cell can be the site of initiation of the disease process. Physicians of Todd's time, however, thought that nerve derangements affecting delivery of the vital spirits to body parts could cause disease, malfunction in the periphery, and alterations in the development of those parts.

Cullen was no longer teaching at Edinburgh when Todd studied medicine there, but Cullen's influence was still dominant in the school of medicine. Cullen began his career as an apprentice in a surgery apothecary; he then became a surgeon on a ship plying between England and the West Indies ˉ ˙ater he joined with William Hunter in an apothecary business in Edinburgh and studied for his medical degree. He became professor of chemistry at Edinburgh, and then of materia medica, and had many brilliant students.

Todd was greatly influenced by Cullen's work and history. Like Cullen, Todd had also served in the navy. While stationed for six or seven years in Italy, he probably performed his newt experiments at the British naval base in Naples. He later returned to England and practiced for many years at Brighton before his death in 1840.

It seems likely that when Todd cut the nerves, he wanted to test experimentally a theory that was then widely accepted as fact, namely, Cullen's theory of the priority of the nervous system. It is true that Todd's experiments were also concerned with other matters of limb regeneration. They included interruption of the vascular supply to the hind limb and a description of the regeneration process itself. Transection of the major artery to the limb did not affect regrowth because, as we now know, collateral circulation is established in the stump within a day. The sciatic artery of the newt is closely associated with the major nerve trunks; even with the modern operating microscope, care must be taken to avoid injuring the sciatic artery when cutting the nerves. Todd makes no mention of using magnifying lenses. He must have performed the operations with the naked eye, a difficult task with so small an animal. We view his separate destruction of the artery as a control for nerve transections, although he does not say this.

Todd's selection of the aquatic newt for his regeneration study was a fortunate one, since the newt's capacity to regrow an appendage was to be, up to the present day, one of the best models for the analysis of the neurotrophic function. Actually, however, Todd was concerned with the broad subject of the regenerative capacity of the limb and tail, and the question of the nervous control of regrowth was a relatively minor aspect of his study. His fascination with regeneration is understandable, since the regrowth of body parts in various animals was a topic of great popular interest in the late eighteenth and early nineteenth centuries, especially in intellectual circles associated with the courts. Figure 7.1 depicts a striking example of the phenomenon in the sequence of regeneration in a forelimb of the newt *Notophthalmus*. During this period appeared the classic studies by Spallanzani, Trembley, Réaumur, and Bonnet on the remarkable ability of certain animals to replace external body parts, including even the head in some species. Indeed, most of Todd's paper simply repeats and confirms Spallanzani's famous observations of fifty years before (1768), with

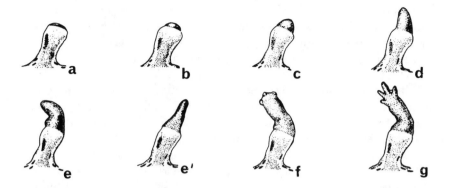

Figure 7.1. Sequence of stages of regeneration of foreleg in the newt *Notophthalmus* based on newts raised at 25° C (magnification about × 5): (*a*) about 10 days after amputation, very early regenerate; (*b*) about 12 days after amputation, early regenerate; (*c*) about 14 days after amputation, medium regenerate; (*d*) about 18 days after amputation; late regenerate; (*e,e'*) about 3–5 weeks after amputation, palette (elbow formation); (*f*) about 6 weeks after amputation, early finger; (*g*) about 8 weeks after amputation, advanced finger. (Reproduced with permission from Singer, 1952.)

which Todd clearly was familiar, although he indicated that he was unable to gain access to a copy himself (Todd, 1823, p. 96).

Studies stimulated by Todd's early work

There have been many studies based on Todd's neurotrophic demonstrations during this century, especially in recent years. These can only be briefly summarized here, but they are described in more detail in various reviews, particularly the extensive ones by Wallace (1981) and Sicard (1985). One of the first studies explored the question of which neurocomponents are responsible for the control of limb regeneration. This question was resolved largely in favor of the view that all components are trophic but that the sensory one has the greatest influence on regeneration. (See Singer, 1952, and Thornton, 1970.) It was also found that the amount of neuroplasm available in the amputation surface is important for regrowth. Nonregenerating limbs of the adult frog, the lizard, the young opossum, and the chick, it was discovered, can be induced to regenerate by increasing the amount of nerve fiber available at the amputation surface. (See Singer, 1954; Mizell and Isaacs, 1970; and Fowler and Sisken, 1982.) There is evidence that the trophic agent is one or more protein substance(s). The search for the neurotrophic substance is now being pursued in a number of laboratories through the use of modern molecular techniques. Interest in the influence of the nerve on the cell cycle continues to increase (Tassava, Laux, and Treece, 1985), although details of how neurons act to evoke cellular responses in the blastema tissue have yet to be uncovered.

These results provide a partial answer to an old question, first posed in scientific literature by the great Italian biologist Spallanzani, whom we are fond of quoting for the clarity of his thought and the beauty of his expression. The ensuing quotation is from his book *Prodromo di un opera da imprimersi sopra la riproduzioni animali* (1768), translated in 1769 as *An Essay on Animal Reproduction*. It is rather disappointing that Todd, in 1823, apparently was unaware of both this translation and the Italian book. He apparently knew of it through hearsay, however, since he repeated some of Spallanzani's experiments on newt regeneration. It should be noted that Spallanzani's book is merely a summary of his biological experiments; the details of his work, described in his unpublished correspondence, have only recently become available. (See Chapter 5 in the present volume.) A sizable portion of Spallanzani's book concerns experiments on regeneration in the salamander, but various comments suggest that he must also have explored the regenerative capacity of other vertebrates. Lizards and frogs were certainly available, neither of which regenerate limbs, although we now know of a few primitive adult frogs that do (Goode, 1967; Michael and Al-Sammak, 1970). Spallanzani was aware, presumably from his experiments, that the frog tadpole, unlike the adult, can regenerate a leg, and he was intrigued that adult frogs, after metamorphosis, do not possess this regenerative capacity. He wrote,

But if these species are able to renew their legs when young, why should they not do the same when farther advanced? . . . Are the wonderful reproductions hitherto mentioned only to be ascribed to the effect of water, in which these animals are kept? This is contradicted in the instance of the salamanders, whose parts were reproduced on dry ground. But if the above mentioned animals, either aquatic or amphibian, recover their legs, even when kept on dry ground, how comes it to pass, that other land animals, at least such as are commonly accounted perfect, and are better known to us, are not endued with the same power? Is it to be hoped they may acquire them by some useful dispositions? And should the flattering expectation of this advantage for ourselves be considered entirely as chimerical? (Spallanzani, 1769, pp. 85–6)

Spallanzani was an experimentalist, and it seems likely to us that he attempted "some useful dispositions" to induce regrowth in nonregenerating animals. For him, the ideal animal to study would have been the adult frog, since early within its life cycle, in the tadpole, the leg does regenerate. The frog was the choice of Thomas Hunt Morgan, whom we know best in his later works as the author of *The Theory of the Gene* (1929). Morgan was concerned with the problem of regeneration early in his career, and in 1901 he wrote a monumental book on the subject. Intrigued by the question of why the regenerative capacity had been lost in vertebrates, he devoted a good portion of an address to the Harvey Society (Morgan, 1906–7) to this problem:

I should like to discuss . . . the question why certain animals seem to lack the power to replace lost parts; and since man himself belongs to this class, the meaning of the fact is of direct and, perhaps, even of practical importance to us; for if we could determine why man does not replace a lost arm or leg, we might possibly go further and discover how such process could be induced by artificial means.

For several years Morgan and his students sought to induce regeneration of the frog's leg by "artificial means." One of his students reported that this could be done by manipulating the skin in relation to the amputation surface (Goldfarb, 1910). This report, however, was merely a note that gave no details of the experimental method, and Morgan himself (1908) reported only unsuccessful attempts.

The parallel decline in quantitative innervation and regenerative capacity in the vertebrate evolutionary series and the demonstrated importance of a "threshold" number of nerve fibers in newt limb regeneration suggested that experimental overloading of the amputation stump with extra nerves might induce regrowth in forms higher than the salamander. The adult frog was again the animal of choice, because it was known that during the tadpole's metamorphosis into the adult frog, there is indeed a decline in nerve number per volume of tissue, and it was theorized that the capacity to regrow, having only recently been lost, might still reside in the limb tissues. The anatomy of the adult frog lends itself well to hyperinnervation experiments, since the frog has long hind limbs, providing adequate length of nerve to redirect into the amputated forelimb (Figure 7.2). Indeed, when we freed the sciatic nerve and its major branches of their investments without interrupting their origins and rerouted them under the flank skin to the stump of the amputated forelimb, the number of nerve fibers at the amputation surface of the upper arm was approximately doubled (Singer, 1954). Induced regeneration occurred in all instances. It was not as perfect as in the salamander, but a small limb, sometimes with fingers, was formed. Induction of frog limb regenera-

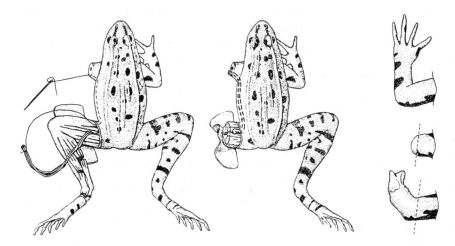

Figure 7.2. Induction of regrowth of frog's forelimb by supplementing the nerve supply to the stump. The sciatic nerve and its branches were dissected free, attached to a thread and needle, and directed under the skin to the amputated forelimb.

The drawings on the right represent (*top*) a normal forelimb; (*middle*) an early regenerate stage at the time of the formation of the blastema; and (*bottom*) the incomplete regenerate, showing two short fingers. (Reproduced with permission from Marcus Singer, "The Regeneration of Body Parts," *Scientific American* 199 [October]: 79–88. Copyright 1958 by Scientific American, Inc. All rights reserved.)

tion using other procedures has also been reported (Polezhaev, 1946). In fact, a "useful disposition," before hyperinnervation, was to traumatize the wound surface repeatedly with hypertonic salt solution (Rose, 1944, 1945). Later analyses showed, however, that without the presence of nerves such insults were ineffective agents of regrowth (Singer, Kamrin, and Ashbaugh, 1957). It is possible that trauma increases the responsiveness of wound tissues to the nerve, causing the limited number of fibers to serve as an adequate threshold.

We were also successful in using hyperinnervation to induce regrowth in the amputated hind limb of the lizard by deviating the sciatic nerve of one side to the stump of the opposite hind limb (Singer, 1961) or by redirecting the caudal spinal cord into the stump (Simpson, 1961). More recently, Fowler and Sisken (1982) obtained similar results in the chick embryo limb. Available mammalian laboratory animals are anatomically less well suited for hyperinnervation experiments than amphibians and reptiles. We attempted to induce regrowth in the mouse hand using two important principles that emerged from our work: hyperinnervation and transplantation. A resected portion of the mouse hand was transplanted to the thigh, near the sciatic nerve. The sciatic nerve was cut and its proximal stump threaded through the transplant to the free surface. We obtained only one instance of regeneration in eighteen successful transplantations (unpublished). The regenerate consisted of a heavily pigmented outgrowth from the end of the stump, starting about one month after the operation. Growth was at first rapid, then stabilized when the total length attained about 2 millimeters. The distal end was ridged and fluted, suggesting rudiments of digits. Histological study revealed skeletal cartilaginous elements, as well as dense connective and glandular tissue, but no muscle.

More consistent regenerative results in mammals were obtained by Mizell (1968), who induced a regenerative response in the amputated leg of the newborn opossum by implanting cerebral tissue (Mizell and Isaacs, 1970). It appears, therefore, that the ability to regrow is latent within the limbs of the higher vertebrates, including the mammals. It is instructive that expression of this capacity can be induced by a morphological manipulation altering the quantitative relation between nerve and other tissues. We should mention here that the possibility of a modest spontaneous regeneration of the digit tips of children and young primates has been reported by Illingsworth (1974) and by Singer et al. (1987).

A tribute to the aquatic salamander

The newt and related species are memorialized in science, fiction, and folklore. They are among the favorite experimental animals because they ask the least of the experimenter. They eat little, heal rapidly, and are subject to few diseases. They are small, timid animals that enjoy each other's company and will sit for hours on a bit of weed with their heads out of water, savoring two environments, as did their ancestors, the first land vertebrates.

To close this discourse we have selected two of many tributes to these animals, one from scientific literature and one from a novel.

From *The Brain of the Tiger Salamander*, by C. Judson Herrick, late professor of neurology at the University of Chicago and one of the great comparative neuroanatomists:

Salamanders are shy little animals, rarely seen and still more rarely heard. If it were not so, there would be no salamanders at all, for they are defenseless creatures, depending on concealment for survival. . . .

This salamander and closely allied species have been found to be so well adapted for a wide range of studies upon fundamental features of growth and differentiation of animal bodies that during the last fifty years there has been more investigation of the structure, development and general physiology of salamanders than has been devoted to any other group of animals except mankind. The reason for this is that experimental studies can be made with these amphibians that are impossible or much more difficult in the case of other animals. This is one justification for the expenditure of so much hard work and money upon the study of the nervous system of these insignificant little creatures.

And from a literary classic by Karel Capek, his satiric novel *War with the Newts,* in which fact and fancy are intertwined:

The last series of the experiments was concerned with the powers of regeneration possessed by the Newts. If a Newt has his tail cut off, a new one will develop in a fortnight; we repeated this experiment seven times with one Newt, in each case obtaining the same result. In a similar manner legs regenerate which have been amputated. We cut off from one experimental animal all four extremities as well as the tail; in thirty days he was complete again. If a Newt femur or scapular bone is broken, the whole limb falls off and a new one grows in its place. It is the same with a damaged eye or a tongue that has been removed; it is a matter of some interest that in the case of a Newt whose tongue I had removed he forgot how to talk, and had to learn again. If one removes the head of a Newt or divides the body between the neck and the pelvis, the animal dies. On the other hand, one can take away the stomach, part of the intestines, two-thirds of the liver, and other organs without injury to life; so that we may state that a Newt that has been eviscerated is still capable of a further existence. No other animal has such resistance to every sort of injury as the Newt. In this respect it might make an excellent, almost indomitable animal for purposes of war; unfortunately, against this stand his peacefulness and natural defenselessness. (p. 198)

ACKNOWLEDGMENTS

We acknowledge with appreciation the years of support we have received for studies on the neurotrophic impulse from the following funding agencies: the American Cancer Society, the Multiple Sclerosis Society, the National Institutes of Health, and the Dysautonomia Foundation. Marcus Singer remembers with gratitude his many colleagues who shared in the work, and the devotion of his assistants.

We are grateful to the Monsanto Company of St. Louis, Missouri, for financial help that enabled us to carry out this present study.

REFERENCES

Brockes, J. P. 1987. The nerve dependence of amphibian limb regeneration. *J. Exp. Biol.,* 132: 79–91.

Capek, K. 1937. *War with the Newts*. First English translation by M. Weatherall and R. Weatherall. London: Allen & Unwin. (Originally published in Czechoslovakian in 1936.)

Drachman, D. B., ed. 1974. *The Trophic Function of the Neuron*. Annals NY Acad. Sci. Press, New York. Vol. 228. Articles by M. Singer, S. Thesleff, D. B. Drachman, and others.

Fowler, I., and B. F. J. Sisken. 1982. Effect of augmentation of nerve supply upon limb regeneration in the chick embryo. *J. Exp. Zool.*, 221: 49–59.

Géraudie, J, R. Nordlander, M. Singer, and J. Singer 1988. Early stages of spinal ganglion formation during tail regeneration in the newt, *Notophthalmus viridescens*. *Am. J. Anat.*, 183: 359–70.

Goldfarb, A. J. 1910. An inquiry into the nature of changes in non-regenerating animals. *Proc. Soc. Exp. Biol. Med.*, 8: 4–5.

Goode, R. P. 1967. The regeneration of limbs in adult anurans. *J. Emb. Exp. Morph.*, 18: 259–67.

Guth, L. 1971. Degeneration and regeneration of taste buds. In *Handbook of Sensory Physiology*. Vol. 4, *Chemical Senses*, pp. 63–74. Springer, Berlin.

Guth, L., ed. 1969. Trophic effects of vertebrate neurons. *Neurosci. Res. Program Bull.*, Vol. 7, MIT, Boston.

Herrick, C. J. 1968. *The Brain of the Tiger Salamander*. University of Chicago Press, Chicago.

Illingworth, C. M. 1974. Trapped fingers and amputated finger tips in children. *J. Pediatr. Surg.*, 9: 409–16.

Michael, M. I., and A. J. Al-Sammak. 1970. Regeneration of limbs in adult *Rana ridibunda ridibunda* Pallas. *Experientia*, 26: 920–1.

Mizell, M. 1968. Limb regeneration: Induction in the newborn opossum. *Science*, 161: 283–6.

Mizell, M., and J. J. Isaacs. 1970. Induced regeneration of hindlimbs in the newborn opossum. *Am. Zool.* 10: 141–56.

Morgan, T. H. 1901. *Regeneration*. Macmillan, New York.

1906–7. The extent and limitations of the power to regenerate in man and other vertebrates. In *Harvey Lectures, 1905–6*, pp. 219–29. Lippincott, Philadelphia.

1908. Experiments in grafting. *Am. Nat.*, 42: 1–11.

1929. *The Theory of the Gene*. Hafner, New York.

Müller, J. 1843. *Elements of Physiology*, 2nd ed. Trans. from German by W. Baly. Lea & Blanchard, Philadelphia.

Parker, G. H. 1932. On the trophic impulse so-called, its rate and nature. *Am. Nat.*, 66: 144–58.

Polezhaev, L. V. 1972. *Organ Regeneration in Animals*. Thomas, Springfield, Ill.

Polezhaev, L. W. 1946. The loss and restoration of regenerative capacity in the limbs of tailless Amphibia. *Biol. Rev.*, 21: 141–7.

Rath, G. 1959. Neural pathology: A pathogenetic concept of the eighteenth and nineteenth centuries. *Bull. Hist. Med.*, 33: 526–41.

Rose, S. M. 1944. Methods of initiating limb regeneration in adult *Anura*. *J. Exp. Zool.*, 95: 149–70.

1945. The effect of NaCl in stimulating regeneration of limbs of frogs. *J. Morphol.*, 77:119–35.

1948. The role of nerves in limb regeneration. *Ann. NY Acad. Sci.*, 49: 818–33.

1970. *Regeneration.* Appleton-Crofts, New York.

Samuels, S. 1860. *Die trophischen Nerven.* Wigand, Leipzig.

Sicard, R. E., ed. 1985. *Regulation of Vertebrate Limb Regeneration.* Oxford University Press, New York.

Simpson, S. B., Jr. 1961. Induction of limb regeneration in the lizard, *Lygosoma laterale*, by augmentation of the nerve supply. *Proc. Soc. Exp. Biol. Med.*, 107: 108–11.

Singer, M. 1952. The influence of the nerve in regeneration of the amphibian extremity. *Q. Rev. Biol.*, 27: 169–200.

1954. Induction of regeneration of the forelimb of the postmetamorphic frog by augmentation of the nerve supply. *J. Exp. Zool.*, 126: 419–71.

1961. Induction of regeneration of body parts in the lizard, *Anolis. Proc. Soc. Exp. Biol. Med.*, 107: 106–8.

Singer, M., R. P. Kamrin, and A. Ashbaugh. 1957. The influence of denervation upon trauma-induced regenerates of the forelimb of the postmetamorphic frog. *J. Exp. Zool.*, 136: 35–52.

Singer, M., R. H. Nordlander, and M. Egar. 1979. Axonal guidance during embryogenesis and regeneration in the spinal cord of the newt: The blueprint hypothesis of neuronal pathway patterning. *J. Comp. Neurol.*, 185: 1–22.

Singer, M.; E. C. Weckesser; J. Géraudie; C. E. Maier; and J. Singer. 1987. Open finger tip healing and replacement after distal amputation in Rhesus monkey with comparison to limb regeneration in lower vertebrates. *Anat. Embryol.*, 177: 29–36.

Spallanzani, L. 1768. *Prodromo di un opera da imprimersi sopra la riproduzioni animali.* Giovanni Montanari, Modena. Translated 1769 by M. Maty, *An Essay on Animal Reproduction.* T. Becket & DeHondt, London.

Tassava, R. A.; D. L. Laux; and D. P. Treece. 1985. The effects of partial and complete denervation on adult newt forelimb blastema cell-cycle parameters. *Differentiation* 29: 121–6.

Thornton, C. S. 1970. Amphibian limb regeneration and its relation to nerves. *Am. Zool.*, 10: 113–18.

Todd, T. J. 1823. On the process of reproduction of the members of the aquatic salamander. *Q. J. Sci., Lit., Arts*, 16: 84–96.

Tomlinson, B. L.; D. J. Goldhamer; P. M. Barger; and R. A. Tassava. 1985. Punctuated cell cycling in the regeneration blastema of urodele amphibians: An hypothesis. *Differentiation*, 28: 195–9.

Vintschgau, M. V., and J. Honigschmied. 1876. Nervus glossopharyngeus and Schmeckbecker. *Arch. Ges. Physiol.*, 14: 443–8.

Wallace, H. 1981. *Vertebrate Limb Regeneration.* Wiley, New York.

Wyburn-Mason, R. 1950. *Trophic Nerves.* Kimpton, London.

8

Regeneration, 1885–1901

FREDERICK B. CHURCHILL

Between 1885 and 1901, regeneration studies in Europe and America increased both in variety and number. To skim through the records of those sixteen years is to discover that droves of experimentalists and anatomists turned to their scalpels, scissors, and heated needles; they shook, ligated, and compressed; they altered the orientation of the egg to gravity and changed the chemical composition of seawater – all with the intent of understanding the "meets and bounds" of regeneration. These biologists – among whom were some of the leaders of experimental biology, such as Wilhelm Roux, Hans Driesch, and Thomas Hunt Morgan – worked with a broad spectrum of organisms and on different levels of organization. They probed eggs, blastomeres, gastrulas, protozoa and algae, and the tissues and organs of adult plants and most common animal phyla for regenerative responses. Pathological material from the clinic and the morgue supplemented material from the laboratory bench. Discussions about neoplasia and hypertrophy, assessments of the regenerative capacities of crystals, and examination of both homoplastic and heteroplastic transplantations helped shape, along with the traditional forms of mechanical assault, their ultimate conclusions. Clearly, one of the problems facing the historian is that of categories. What was considered regeneration? What was not? Is there a natural divide between regeneration experiments, on the one hand, and experimental embryology, on the other?

I intend to present a selective survey of this apparent chaos of activity. I draw upon a decade of review articles (1891–1901) by the Rostock anatomist Dietrich Barfurth on the subject of "Regeneration and Involution" as my primary source of information. Barfurth studied medicine, and in 1883 habilitated in anatomy, at the University of Bonn. In 1888 he became prosector of anatomy under Friedrich Merkel in Göttingen; in 1889 he attained the rank of Ordinarius of comparative anatomy, histology, and embryology at the University of

113

Dorpat. It was during his tenure in this position that he wrote the majority of the reviews I survey. In 1896 he became director of the anatomical institute in Rostock, where he spent the rest of his scientific career.[1]

Twenty-three of these reviews appeared in Merkel and Bonnet's *Ergebnisse der Anatomie und Entwickelungsgeschichte* between 1891, when the journal began, and 1916, when it came to an end. The reviews consist of lengthy commentaries on the work done on regeneration, and to a much smaller extent on degeneration, or what Barfurth called *Involution*. In brief, these annual reviews represent a genre of scientific literature that was common and necessary in the nineteenth century. What seems unusual in this case is that a single author wrote the entire series, from the first volume of the *Ergebnisse* to the last before the journal was suspended in the middle of World War I. Here is an opportunity to get a general picture of the early development of a well-defined field of study. In this essay I examine only the first ten years of Barfurth's redactions and evaluations.[2] It is important to realize that Barfurth was not an impartial commentator. He performed his own extensive regeneration experiments on amphibian limbs and was almost an exact contemporary and an admirer of Wilhelm Roux. Because he represents the "losing side" on what was identified as Roux's "mosaic" theory of development, Barfurth's evaluation of events is an unfamiliar one and a good antidote to traditional histories of the period. In addition, I intend to anchor my ten-year survey to two familiar landmarks, one on each side of the decade. I choose Paul Fraisse's monograph appositely entitled *Regeneration*, which appeared in 1885, and Thomas Hunt Morgan's book with a similar title, which appeared in 1901.[3]

Before beginning the survey, however, I shall make several generalizations about the traditional accounts of regeneration research during this period. First, they tend to be distorted by what I shall call

---

[1] For an autobiographical sketch of Barfurth, see *Die Medizin der Gegenwart in Selbstdarstellungen*, ed. L. R. Grote (Leipzig, Meiner, 1923), pp. 1–22. See also Magnus Schmid, "Barfurth, Dietrich, " in *Neue deutsche Biographie* (Berlin: Duncker & Humboldt, 1952–), 1:588.

[2] I examined the entire sequence of reviews with another set of questions in mind in "Regeneration in Its Heyday: 1891–1916," an address presented at the annual meeting of the History of Science Society, Gainesville, Fla., October 28, 1989.

[3] Paul Fraisse, *Die Regeneration von Geweben und Organen bei den Wirbelthieren, besonders Amphibien und Reptilien* (Cassel: Theodor Fischer, 1885), and Thomas Hunt Morgan, *Regeneration* (New York: Macmillan, 1901).

"Gipfelsammler's myopia."[4] This dread disease commonly afflicts historians of science, philosophy, art, and other areas of high culture. Its new name parodies an alpinist's term that refers to a single-minded attention to dramatic mountain peaks accompanied by total neglect of the surrounding hills and valleys that lend definition and meaning to the territory. Thus in our histories we normally read of the counterpoint between Roux's famous 1888 half-embryo experiments on frogs and Hans Driesch's equally famous 1891 shaking experiments on sea urchin eggs. We find mention of Gustav Wolff's ablation of the lens from the eye of salamanders in 1894, Curt Herbst's turn-of-the-century discovery that calcium-free seawater caused the blastomeres of sea urchin eggs to separate, and Hans Spemann's constriction experiments on salamander eggs and lens induction experiments on frogs in 1901.[5] I do not wish to denigrate these events or the historians who have sought to understand their details, but to focus on them alone is to give the impression of a few investigators' making fabulous discoveries in a wilderness of scientific indifference. This view I wish to challenge.

Second, I should like to guard against the "crucial experiment fallacy," which creeps into our histories and is often a consequence of Gipfelsammler's myopia. For example, it is not uncommon to find Driesch's 1891 experiment presented as a decisive refutation of the conclusions that Roux drew from his 1888 experiment. The ensuing survey will show how far off the mark that judgment is. Experimental embryology abounds with purported *experimenta crucis*. Ross G. Harrison's 1907 experimental production of nerve fibers from isolated

---

4 The German expression *Gipfelsammler* literally denotes an alpinist who climbs mountain peaks solely to record them in his records. Just like the historian who commits this error, the climber focuses only on the peaks and ignores the magnificent scenery surrounding them.

5 Detailed accounts of these experiments can be found in Jane Oppenheimer, "Questions Posed by Classical Descriptive and Experimental Embryology," in Oppenheimer, *Essays in the History of Embryology and Biology* (Cambridge, Mass.: MIT Press, 1967), pp. 62–91; Frederick B. Churchill, "Chabry, Roux, and the Experimental Method in Nineteenth-century Embryology," in Ronald N. Giere and Richard S. Westfall, eds., *Foundations of Scientific Method* (Bloomington: Indiana University Press, 1973), pp. 167–205; Churchill, "From Machine-Theory to Entelechy: Two Studies in Developmental Teleology," *J. Hist. Biol.* (1969) 2:165–85; T. J. Horder and Paul J. Weindling, "Hans Spemann and the Organiser," in T. J. Horder, J. A. Witkowski, and C. C. Wylie, eds., *A History of Embryology* (Cambridge: Cambridge University Press, 1986), pp. 183–242; and Viktor Hamburger, *The Heritage of Experimental Embryology: Hans Spemann and the Organizer* (New York: Oxford University Press, 1988).

frog larvae neuroblasts is commonly described as such.[6] Spemann's and Hilde Mangold's production of a secondary embryo by implanting the dorsal lip of the blastopore seemed decisive at the time, but no one could figure out what it was that it had decisively demonstrated. That is just the problem. A crucial experiment implies that an act of manipulation can be decisive in choosing between two competing theories. It asserts that an empirical result will logically prove or disprove a theory. But theories are not proved or disproved in Euclidian fashion. They are only confirmed or not confirmed, and, like old soldiers, the poorly confirmed ones do not die; they simply fade away. Moreover, experiments are not simply acts of manipulation. They are complex acts, components of which are in themselves theory-laden. They are often the summation of individual manipulations in which the outcome is often a spectrum of results that must be analyzed statistically. In his study of Harrison, Witkowski generalizes on this point. "Experiment in itself was not sufficient. Experimental analysis involved an understanding of the problem, the framing of questions that were both interesting and capable of being answered, and the design of experiments that provided answers that were unequivocal if possible."[7] We might conclude that manipulation is accompanied by examination, interrogation, and consideration – all of which provide plenty of room for equivocation. This is hardly a formula for a crucial experiment.

Paul Fraisse, 1885

Having fortified ourselves against two malaises afflicting the history of science, let us turn to our anchor point in the mid-1880s, that is, to Paul Fraisse's monograph. I can discover little about Fraisse's own career. When he wrote *Regeneration* in 1885, he was a doctor of medicine and philosophy and a privatdozent at the University of Leipzig. In addition to this important monograph, he wrote smaller contributions on fish cultivation and mollusk development. He may not have fared well at Leipzig, for between 1896 and 1909, when he died, he is described in *Minerva* as an *Ordinarius* in zoology but at the same time as *beurlaubt* – that is, on leave of absence. *Regeneration* was a result of

[6] On Harrison, see Jane Oppenheimer, "Ross Harrison's Contributions to Experimental Embryology," in *Essays*, pp. 92–116. For a different perspective, see Jane Maienschein, "Experimental Biology in Transition: Harrison's Embryology, 1895–1910," *Studies in the History of Biology* (1983) 6:107–27.

[7] Jan A. Witkowski, "Ross Harrison and the Experimental Analysis of Nerve Growth: The Origins of Tissue Culture," in Horder, Witkowski, and Wylie, *History of Embryology*, pp. 149–77.

six years of investigation into the regenerative capacities of various organ systems of amphibians and reptiles.

His concerns were those of the immediate post-Darwinian and post-Virchowian period. Fraisse set about establishing a taxonomy of regenerative capacities. He excised various sections of epidermis and cutis, of subjacent epidermal glands and organs, of the skeletal system, vertebrae, and nervous system, and of the muscle and vascular systems. He correlated his findings concerning an animal's regenerative response with the animal's health, its age, and the season. He found his endeavors speaking directly to questions about phylogenetic relationships and to the Virchowian perspective that viewed the organism as a cell state in which each cell was derived from earlier cells through direct cell division and in which the cells in general formed the basic unit for understanding the functions of life. Fraisse also suggested that Roux's recent extension of natural selection to tissues and cells allowed the biologist to understand formation, reformation, and malformation on a new level.[8]

Fraisse addressed five general issues throughout his account: (1) he was familiar with the most recent cytological studies demonstrating nuclear continuity in normal mitosis, but Fraisse could not bring himself to reject the possibility of free nucleus and free cell formation at the epithelial edges of regenerating wounds and amputated parts. He argued that even Walther Flemming, the most forceful proponent of the doctrine *Omnis nucleus e nucleo,* or "every nucleus from a nucleus," was not fully convinced that the process of regeneration was an exception to his fiat.[9] (2) Fraisse also felt a case could be made for direct, as opposed to indirect or mitotic, cell division during the regeneration of certain tissues.[10] (3) He recognized that regeneration studies spoke also to the issue of the origin of the various tissue elements and (4) to contemporary debates over the homology of the germ-layer doctrine. In both cases he felt that in higher organisms regeneration supported the argument for specificity of germ layers and tissues. The connective tissue, he argued, however, might produce tissues of widely different kinds.[11] (5) Finally, Fraisse was interested in the relationship between the regenerative process and ontogeny and phylogeny. He recognized that regeneration sometimes repeated normal develop-

---

[8] Fraisse, *Regeneration*, pp. iii–viii. Wilhelm Roux, *Der Kampf der Theile im Organismus. Ein Beitrag zur Vervollständigung der mechanischen Zweckmassigkeitslehre* (Leipzig: Wilhelm Englemann, 1881).
[9] Fraisse, *Regeneration*, pp. 55–6, 140–1.
[10] Ibid., pp. 141–4.
[11] Ibid., pp. 141–6.

ment, but he was inclined to argue that regeneration was "neither a pure recapitulation of ontogenetic or phylogenetic developmental processes nor [was] it alone explainable through the relationship of functional adaptation." Instead, Fraisse invoked, without explaining either, a combination of hereditary and correlative developmental events.[12] His monograph was not the only work of the decade on regeneration, but it was often cited and highly praised in later accounts.

Fraisse completed his monograph in 1885. It represented not only the culmination of his half-dozen years of investigation of regeneration but also the waning concerns of several post-Darwinian and post-Virchowian traditions in experimental morphology. The same concerns were expressed, to varying degrees and with different emphases, in the contemporary works of Camille Dareste, Laurent Marie Chabry, Wilhelm His, Leo Gerlach, and August Rauber.[13] It would be wrong to imply that these investigators shared a common research tradition; they certainly did not. None of them, however, tried to emulate the experimental tradition in physiology that had come to maturity during the previous thirty years. This is the motivation, however, that lay behind the next wave of regeneration studies.

In 1883 the Bonn physiologist Eduard Pflüger published an effort to correlate by carefully controlled experiments what he considered the initial structure of the fertilized egg, the planes of cleavage, and the orientation of the primary axes of the embryo.[14] Within a twelve-month period, three morphologists – Wilhelm Roux, Gustav Born, and Oscar Hertwig – rose to the bait. They disagreed with Pflüger's conclusions, which seemed far too simple to explain the complex pro-

---

[12] Ibid., pp. 147–51 (quotation on p. 154). Note that Fraisse uses the expression *Vererbungsercheinung.*

[13] The major works of these biologists were: Camille Dareste, *Recherches sur la production artificielle des monstruosités ou essais de tératogénie expérimentale* (Paris: C. Reinwald, 1877); Laurent Marie Chabry, "Contributions a l'embryologie normale et tératologique des ascidies simple," *Journal de l'anatomie et de la physiologie normales et pathologiques de l'homme et des animaux* (1887) 23:167–321; Wilhelm His, *Unsere Körperform und das physiologische Problem ihrer Entstehung. Briefe an einen befreundeten Naturforscher* (Leipzig: F. C. W. Vogel, 1874); Leo Gerlach, *Die Entstehungsweise der Doppel Missbildungen bei den höheren Wirbelthieren* (Stuttgart: Ferdinand Enke, 1882); and August Rauber, *Formbildung und Formstörung in der Entwickelung von Wirbelthieren* (Leipzig: Wilhelm Engelmann, 1880). For discussions of these works, see Frederick B. Churchill, "Wilhelm Roux and a Program for Embryology," Ph.D. diss., Harvard University, 1966; Churchill, "Chabry, Roux."

[14] Eduard Pflüger, "Über den Einfluss der Schwerkraft auf die Theilung der Zellen," *Archiv f. d. gesammte Physiologie des Menschen und der Thiere,* pts. 1 and 2 (1883) 31:311–18; 32:1–79.

cess of the development of organic form. In their rebuttals, however, the three anatomists by training were forced to devise experiments of their own that satisfied the physiologist's demands for an analysis of isolated proximal factors.[15] From then on, regeneration studies had to measure up to these exacting methodological requirements. Barfurth rode the new challenge into the next decade.

Dietrich Barfurth, 1891–1901

In Barfurth's reviews, we immediately discern a heightened level of professional involvement. This is easily detected in passages that convey a sense of the excitement inspired by regeneration experiments, particularly on the egg and blastomeres. For example, after completing his review of the year's work for 1892, Barfurth mentioned that he had just received August Weismann's *Keimplasma,* which he promised to review the following year. "The book," he reported, "contains among other things a complete theory of regeneration, which for our area should mark an outstanding literary event."[16] The following year Barfurth noted that the reviewed literature "contains more controversial broadsides [*Kampfschriften*], than one might expect in the space of a chapter and in such a small timespan. This proves," he continued, "an unusual interest on the part of investigators for the emerged problems and can only advance our understanding" (3:175). Barfurth opened his review of 1894 with mention of a controversy being "waged" over "direct and indirect" development, "which is pursued like a real border dispute with enmity and stubbornness and not always with drawing-room decor." "But this battle," he again insisted, "has the benefit that it brings us a deeper insight into the way and capacity of regeneration than was permitted to the older genial experimenters Spallanzani and Bonnet (4: 463). The next year Barfurth described the ever increasing interest of researchers in regeneration studies but cautioned that "we are still very far from a true explanation of regeneration and . . . the many new facts actually present us

---

15 For an account of the experiments of this period (1882–4), see Thomas Hunt Morgan, *The Development of the Frog's Egg: An Introduction to Experimental Embryology* (New York: Macmillan, 1897).

16 Dietrich Barfurth, "Regeneration," *Ergebnisse der Anatomie und Entwickelungsgeschichte* 2 (1892): 156. In the following years the annual review was entitled "Regeneration and Involution." Citations, hereafter given in parentheses in the text, include date of year under review (in some cases), volume number, and page number. As with all annual reviews, the publication date normally follows the year under review but is not associated with the standard citation. All translations are by the author.

with ever new puzzles" (5: 368).[17] In reviewing a half decade of Driesch's experiments in 1896, Barfurth confessed to "trembling with fear," and he closed his report with a gesture to the reader who has the impression "that the number of those who wish to bake in the great kitchen has again increased" (6: 400 and 425).[18] By 1897 it was clear that the controversy had become ugly. Barfurth spoke of a *Hatz*, or hounding, of Roux by an opposition that "sought to silence him and to harangue him to death with speech and broadside" (7: 495).[19] This tempest subsided by the end of the decade, but the interest in regeneration did not. Writing on the literature for 1899, Barfurth spoke of a bumper crop of fine regeneration experiments from all cultivated nations (9: 406).[20]

It is worthwhile making two observations on this string of comments. First, as any reader of Roux's or Oscar Hertwig's contributions knows, these new regeneration studies were loaded with implications for conflicting theories of development and what one might call "territorial disputes" over leadership of the newly emerging field of experimental embryology. Second, and more relevant to my purposes, is the rapid expansion of regeneration experiments, both in range of experimental organisms and in number of investigators. By all signs, regeneration studies were an exciting area in experimental biology.

Research areas and problems

Barfurth included bibliographies of between one hundred and two hundred items in his annual reviews.[21] He commented on only a portion of this work, but he surveyed a range that included studies on inorganic crystals, single-cell organisms, eggs and blastomeres, germ

[17] He also mentions here that for a while the most important effort should be to collect more data and that only this would permit researchers to get closer to an appropriate theory of regeneration.

[18] "Mit Zittern und Zagen gehe ich in diesem Jahre an die Besprechung des in den Überschrift charakterisierten Kapitels."

[19] "Roux totzuschweigen, totzureden und totzuschreiben."

[20] "Dass Berichtsjahr (1899) brachte eine erstaunliche Zahl guter experimenteller Untersuchungen aus allen Gebieten der Regenerationsforschung. Und an diesen Leistung sind alle Kulturnationen beteiligt!" – "The year under review (1899) issued forth an astonishing number of good experimental investigations from all areas of regeneration research. And all cultured nations took part in this accomplishment."

[21] Barfurth's annual bibliographies, which precede the essay reviews, are not consistently drawn up so as to provide an easy way to construct graphs showing the increase in interests. His comments, however, document the heightened involvement of investigators in regeneration research.

layers, adult tissues, and human pathological and surgical material. He also described attempts to fuse embryos and referred to botanical investigations. I limit my examination to the first four of these categories. For reasons that will immediately become apparent I start with the second.

Protozoa

One area of research that attracted a good deal of attention from the very beginning was the regeneration of experimentally divided Protozoa. I was surprised to learn that Ernst Haeckel and Richard Greeff performed regeneration experiments on single-cell organisms in the 1860s. Between 1884 and 1887, Moritz Nussbaum and Weismann's brother-in-law August Gruber performed an extensive series of experiments on Protozoa, studying the ciliates *Oxyticha* and *Stentor*. By 1891, Edouard Balbiani, Max Verworn, and Bruno Hofer had performed additional experiments. Barfurth felt it was important to survey all of this pre-1891 work in his first review (1: 113–15). He often returned to the subject in later years.

Several matters are worth noting about this line of research. First, these experiments on Protozoa commenced within a year of the time when Roux, Hertwig, and Gustav Born began their own experiments in response to Eduard Pflüger's reorientation experiments on frog eggs and so, tradition has it, initiated modern experimental embryology. What is more, these protozoologists were asking questions about the relationship between the cell nucleus and the cell protoplasm at least four years before Roux attempted his hot-needle experiment, which, in a cruder way, asked the same questions. It seems to me that Nussbaum and Gruber deserve a share of the credit for directing zoology into the domain of experimental morphology. Second, it is important to realize that it was only in the mid-1870s that Protozoa were accepted as single-celled organisms. This identification was essential before a comparison between the fission of Protozoa and the karyokinetic activities in embryonic and adult metazoan cells could be made.[22] Nussbaum, Barfurth reported in 1891, claimed his experiments demonstrated that both a nucleus and protoplasm are neces-

---

[22] For historical accounts of the establishment of the cellular analogy with Protozoa and an assessment of its long-term implications, see Frederick B. Churchill, "The Guts of the Matter: Infusoria from Ehrenberg to Bütschli: 1838–1876," *J. Hist. Biol.* (1989) 22: 189–213, and Natasha X. Jacobs, "From Unit to Unity: Protozoology, Cell Theory, and the New Concept of Life," *J. Hist. Biol.* (1989) 22: 215–42.

sary for regeneration (1: 115). After initially disagreeing with him, Gruber also supported this position. Balbiani refined these conclusions by identifying regeneration in ciliates with the macro- rather than the micronucleus. In his review for 1892, Barfurth explained that Balbiani, by assigning the developmental and regenerative functions to different subcellular locations, "had established a new basis for the entire theory of regeneration" (2: 134). By 1894, reports on this line of experiments disappear from Barfurth's reviews. Protozoa appear again briefly in the review of the literature for 1897, but then in the context of searching for the minimal size that a fragment of egg or of a protozoan must have in order to permit regeneration (7: 495).

Inorganic crystals

A second area of research concerned a comparison between organic regeneration and crystal formation. Herbert Spencer, whose *Principles of Biology* had been translated into German in 1876, had suggested that the formation of crystals is analogous to regeneration in organisms (1893, 3: 177). Barfurth's first reference to this subfield of investigation came in 1891 in his first review. In 1893 he reported that Wilhelm Haacke had based an entire book on the analogy and had argued that this comparison showed that "regeneration is . . . completely self-explanatory" (3: 178).[23] Beginning with his review of the 1895 literature, Barfurth devoted an independent section to regeneration experiments on inorganic crystals. In 1895 he mentioned six different investigators of such phenomena, the most important of whom was the Dorpat anatomist August Rauber. Rauber had earlier experimented on what he called "cellular mechanics," or the study of the size, shapes, and movements of cells (Barfurth, 1895, 5: 337–41).[24] By middecade Rauber turned his attention to a microscopic analysis of the fractured edges of broken crystals and to the ensuing patterns of crystal formation. For the remainder of the decade Rauber published yearly accounts of further experiments, each with an accompanying atlas of photographs. By the time of his 1900 report, Barfurth could conclude that, "with the basic industry of the German scholar," Rauber had demonstrated that the geometri-

---

[23] Haacke's statement continues, "When a salamander regenerates a lost limb or a lizard a broken-off tail, the newly appearing cells must necessarily so organize themselves in a way, as has been prescribed to them by the configuration of the still present cells and the form of their Gemmarien [or plasmic particles]."

[24] August Rauber, *Formbildung und Formstörung* (1880).

cally precise repair of crystals could not be mistaken for the far more plastic regeneration processes found in the organic world (10: 585–6). Despite what Spencer, Haacke, and others had asserted, a crude reductionism was not going to explain organic regeneration.

Eggs and blastomeres

Far more dominant in Barfurth's redactions were the experiments performed on eggs and early embryonic cleavage. Many of the colorful quotations that illustrated the excitement and passions of regeneration studies are drawn from this subfield of research. Space does not permit me to do justice to the ramifications of individual experiments of the period. Besides the famous experiments by Roux in 1888 and Driesch in 1891, there were scores of less heralded investigations that added insights, complications, and new twists to the responses of altered or isolated blastomeres. Singling out Barfurth's review of the literature in 1895 as a sample, we can find discussions of: (1) shaking experiments on eggs of two species of sea urchin by T. H. Morgan; (2) more experiments on the prospective significance of urchin blastomeres by Driesch; (3) reexamination of Chun's experiments on ctenophores by both Morgan and Driesch, working together; (4) examination of the first two blastomeres of salamander eggs by Amedeo Herlitzka; (5) studies on two species of *Rana* by H. Endres and H. E. Walter, working in Weismann's institute; (6) ligature experiments on salamander eggs by Endres; (7) Chun reflecting on Morgan and Driesch's criticisms; (8) Driesch countering Chun; (9) four papers by Roux reflecting on his own experiments and those of his critics; (10) half-embryo experiments on frogs by Morgan; (11) isolation experiments on *Medusa* by R. Zojo; (12) experiments on the eight-cell stage of frog embryos by P. Samassa; (13) a highly theoretical analysis of development and regeneration by Driesch; and (14) the first highly polemical monograph entitled *Zeit- und Streitfragen der Biologie* by Oscar Hertwig.

This list is drawn from the work reported for only one season. A fraction of all this literature appears in the traditional historical accounts; most of it does not. All of it, however, forms the texture of the investigations into regeneration; collectively the papers tell a different story from that told by those single premier experiments that customarily get passed to future generations as part of the lore of science past. Seeing it all laid out in annual reviews reveals instead an advancing research front that is self-supporting, self-generating, and interactive. There are momentary triumphs, short-term gains, and setbacks. There are also outstanding investigators, such as Roux, Driesch, and Morgan, who guide the bulk of the research, but the irony is that the extent of their involvement and influence becomes clear only when the entire front is visualized. How does the historian deal with this superabundance of technical, and often contentious, literature from even a single decade?

Short of despairing, let me venture some generalizations gleaned from only a partial understanding of what was going on.

First, this field presents a classic example of an expanding science. There was no consensus about experimental results. Different groups of organisms seemed to render different outcomes; even different species within the same genus could differ in response in significant ways. Responses also appeared to be highly dependent on the specific technique employed and on the stage of development of the organism. This sounds a little like Kuhn's description of "science in crisis," but that expression connotes despair rather than the open-ended excitement and forward momentum that seems evident in Barfurth's reviews.

Second, it seems clear from Barfurth's account that one cannot automatically separate investigators and their results into two neatly isolated and antithetical camps, as the authors of later accounts are wont to do. True, Barfurth often designated a Roux-Weismann position, which he identified with the notions of *Mosaikarbeit* and qualitatively unequal nuclear division, and he counterbalanced this with a Driesch-Hertwig position, which he identified with notions of *regulation* within an equipotential system. At other times, however, there appeared to be little unanimity within camps and results. Weismann did not accept Roux's account of *Postgeneration* until the end of the decade; Roux and Barfurth joined Morgan in attacking Weismann's view of regeneration as an adaptive trait (1899, 9: 409–10). Morgan was critical of the Driesch-Hertwig concept of "interaction of the cells" as well as the Roux-Weismann theory (1895, 5: 347). Morgan, Driesch, Fischel, and Chun could not even agree as to what constituted half- and quarter-ctenophore larvae (1895, 5: 354b; 1897, 7: 498). Driesch recognized an increasing specification of cells after third cleavage in sea urchins (1896, 6: 402), and both Driesch, with sea urchins, and Morgan, with frogs, obtained "accidental, individual variations" that ranged from half, normal-sized embryos to whole, small embryos (1896, 6: 402). When the anatomist Carl Rabl wrote in 1899 a review of the various experiments on blastomeres, he reported that they "bring us up against one of the most difficult, yes perhaps the single most difficult problem with which experimental embryology has presented us up to date" (quoted in Barfurth, 1899, 9: 349). As historians, we must recognize that there were many levels of disagreement and temporary alliances as regeneration studies advanced during the decade. What is clear to us in retrospect consisted of an exciting confusion of multiple results at the time. The "crucial experiment" is more an artifact of retrospective stocktaking than a consequence of a "eureka" experience.

Third, one major source of disagreement and confusion was the definition that Roux and Driesch gave to the term "regeneration." In his famous half-embryo paper, Roux devoted most of his essay to a discussion of what happened to the operated blastomere(s). It is symptomatic of how historians have treated Roux's account that the fine recent English translation omits this portion of the text.[25] For Roux, this would have been unforgivable because it was here that he first introduced the notion of "postgeneration," that is, a reorganization of the protoplasm and a reconstruction of the missing half of the embryo. Later Roux considerably sharpened his definition of the term. He did not wish to consider this process similar to the regeneration associated with the loss of an adult limb or tail, for the obvious reason that the second half of the embryo had not been lost, in the literal sense. On the other hand, Roux did not want the process to be confused with the normal development that took place in the unaltered blastomere. He therefore wrote about "direct" (i.e., normal) and "indirect" (i.e., abnormal) development. Barfurth was emphatic about the potential confusion: "[Roux] is guilty on this score, now that the whole world shakes eggs, scarcely one of ten anatomists – and even this one perhaps without complete confidence – knows where development stops and regeneration begins." (1892, 2:129). Hertwig, Driesch, and Morgan would not accept postgeneration as a distinct phenomenon. They considered most of the alleged cases to be examples of regulation within an equipotential system. (See, for example, Barfurth, 1899, 9: 352.)

If the term "postgeneration" confused many investigators, Driesch refused to consider the reordering of new cells during the closure of what appeared to be half *Echinus* blastulas as a form of regeneration. Barfurth felt that this denied by an inordinately restrictive definition the possibility of self-differentiation of the regenerating part (1900, 10: 557–8).

Fourth, what was at stake in the many arguments concerning the meanings of postgeneration, regeneration, and normal development

---

[25] See Benjamin H. Willier and Jane M. Oppenheimer, *Foundations of Experimental Embryology* (Englewood Cliffs, N.J.: Prentice-Hall, 1964), pp. 2–37.

was a basic model of development that seemed to dictate how the investigator read the experimentally induced response. I shall say more about this in a moment.

Experiments on germ layers and tissues

The last area of experimentation I wish to mention comprised the work on embryonic germ layers and tissues. In sheer bulk of publications, this represented far more effort than the other three areas that I have mentioned so far. Barfurth's first report reviews for a second time the work of Moritz Nussbaum. In an ingenious replay of one of Trembley's classical experiments on hydra, Nussbaum provided a new way of examining the regenerative roles of the ectoderm and endoderm. Weismann's student Chiyomatsu Ishikawa tested and elaborated the procedure, which (briefly) consisted of a mechanical eversion of the hydra followed by the use of physical restraints to prevent it from quickly returning to the normal configuration. Would the two layers of the hydra exchange functions? Both investigators discovered that the hydra, if it were physically possible, found ways of reestablishing the normal relationships even when it meant tearing and dissociating the internally located ectoderm. If it could not manage the act, the animal soon died. From this followed strong assertions about the specificity of the germ layers and Nussbaum's suggestion that embryonic cells in the mesoglea might participate in the regenerative process.[26]

By the early 1890s, biologists had explored the regenerative capacity of a whole range of animals. Barfurth worked on frog larvae, Arnold Lang, F. von Wagner, and F. Braem on hydroids, Harriet Randolph on earthworms. These were followed later in the decade by L. S. Schultze on *Ciona;* E. Schultze on polychaetes; E. Schultze and Driesch on tubularians; and Morgan on an array of organisms but, most important, on planarians. Many of Morgan's students, including Florence Peebles, Helen King, Johanna Kroeber, Annah Hazen (I assume these student scientists were all from Bryn Mawr), published their own regeneration experiments. This is simply the tip of the iceberg. Most of these experiments were variations of amputations or incisions that stimulated the regeneration of a lost or partly destroyed body part. Although one can find disagreement on the interpretation of cellular origins and movements and although there were some ambiguous situations, such as with the pharyngeal region of *Planaria* and the caudal region of *Lumbriculus,* by the end of the century there

---

[26] Quotation appears in Barfurth, 1893, 3:161.

seemed to be a general reconfirmation of Fraisse's assertions that in regeneration ectodermal and endodermal tissues reproduced according to kind. It was not only Barfurth and other followers of Roux's concept of *Mosaikarbeit* in normal development but also epigeneticists such as Driesch who adopted this position for regeneration beyond the cleavage level. In 1900 the implications of much of the regeneration work seemed to be a weak endorsement for a belief in the specificity of the germ layers and more advanced tissues, but this important embryological issue was not to be quickly decided.[27]

One set of experiments, in particular, failed to fit easily into the germ-layer model of development. In 1894 Gustav Wolff experimented with the regeneration of the lens following its extirpation from the salamander eye. Similar experiments had been performed earlier by Emery and Vincenzo Colucci. Wolff, however, showed that the lens did not regenerate from surface ectoderm, as it normally develops, but from the iris of the eye. The surprising results led Wolff to renounce a mechanistic explanation (1894, 4: 488), but other workers pursued the phenomenon without despairing. Immediately Erik Müller and W. Koch, and later in the decade Fischel, Brachet, and Benoit, confirmed and added more information about this regenerative process. Throughout the period Barfurth remained puzzled by the results. One commentator compared the phenomenon to a loose parade horse charging off in all directions – against "the theory of descent here, the doctrine of natural selection there, and naturally against the germ layers and the specificity of the tissues." Fischel offered some solace by suggesting that the iris, after all, was derived from the neural ectoderm and therefore indirectly came from the same embryonic germ layer as the original lens (1900, 10: 589; quotation on p. 575).

Models of development

I have yet to write of regeneration experiments in botany and in pathology, or of the fusion of two embryos and transplantation of tissues between organisms, all of which are discussed by Barfurth. The decade abounded with such work. Despite these historical omissions, what can one say of general interest about the investigations I have already described?

First, I hope this partial account has made it clear how extensive and varied the research into regeneration was in the 1890s. There were many ambiguous results and competing explanations for the particu-

---

[27] This weak endorsement disappears after the turn of the century.

lars. One may well imagine that this would normally be true of a rapidly developing area of science – a "robust science," as we might call it. We may also imagine that when a number of disciplines – in this case, physiology, embryology, protozoology, botany, pathology, and crystallography – converge on a set of perceived common phenomena, there will be considerable disagreement in the ranks about definitions, methods, and conclusions. That Roux and Driesch could not even agree about what constituted regeneration in the dividing egg seems to me symptomatic of this state of simultaneous convergence and rapid expansion.

Second, I have presented a single window onto this research, created by Dietrich Barfurth, who himself was deeply involved in regeneration experiments and who openly and repeatedly supported Roux's interpretation of the results. There are advantages to this, for it allows us to appreciate that what in some measure would later appear as the "losing side" was really, throughout the decade, a thriving and developing position. Barfurth, in fact, was confident that the results confirmed Roux's concepts of *Mosaikarbeit* and postgeneration. The doctrine of specificity of the germ layers and tissues seemed more securely documented than in Fraisse's day, and both Roux's contrast between self-differentiation and dependent differentiation and his contrast between normal and abnormal development seemed useful for analyzing regeneration. Roux's concept of postgeneration seemed an important means for discussing experimental results on the early stages of embryonic development. Although Barfurth was never a supporter of Weismann's germ-plasm theory, he endorsed the Roux-Weismann theory of qualitatively unequal nuclear division and the idea of a reserve idioplasm of some nature. Many others did, too. Finally, it is worth pointing out that Barfurth never identified Roux, Weismann, or himself as a "preformationist." This was a polemical term resurrected, I suspect, by Hertwig. Barfurth, and perhaps most other German scientists, used instead another embryological term: "evolutionist."

Driesch must rank with Roux as a leading commentator on regeneration before 1901. Many of his concepts, such as regulatory development and a harmonious equipotential system, were important influences on embryology. Although he was the initial challenger of Roux's half-embryo experiments, his opposition to Roux seems quite benign, and his regeneration experiments on more advanced stages of development supported Roux's notions of a *Mosaikarbeit* and of the specificity of the germ layers, which is seldom mentioned in histories that wish to emphasize the dramatic contrast between their famous blastomere experiments. To my mind, it is more meaningful to emphasize the contrast between Driesch's neo-Kantian ontology and Roux's materialism.[28] The vitalistic metaphysics to which Driesch later turned did not destroy a useful dialogue between the two experimenters. If I rightly remember, Driesch used to stay at Roux's home when he passed through Halle.

I am less familiar with Hertwig's experiments during this period. Barfurth's reporting of Hertwig's polemical assessments make this decade appear as tumultuous as it was creative. I have read Hertwig's broadsides in other contexts and have always considered them too negative to be useful. Many of his contemporaries clearly thought so as well. Until the turn of the century, Barfurth continued to insist that regeneration of differentiated tissue presented the strongest case against the Driesch-Hertwig doctrine of qualitatively equal cell division. After all, if the content of all cells were similar, the Driesch-Hertwig doctrine of cell totipotency would require auxiliary hypotheses to explain the specification of cell and tissue types. This was, of course, the same argument in reverse that Hertwig and others used against any theory of reserve idioplasm to explain regeneration.

With regard to these polemics, one Hertwig comment in particular, reported by Barfurth, caught my eye: "The process of the division of labor and differentiation, as it takes place between the cells of an organism, rests on other grounds than on the process of species formation in the animal and plant kingdom and is perhaps to be compared in its essence to the analogous process in human society" (1898, 8: 674). It takes a moment to realize what Hertwig was driving at with this set of metaphors. He rightly recognized that embryologists in the nineteenth century, from the time of von Baer, had misconstrued differentiation as a formal process of "increasing individualization." This most fundamental principle appeared as the overall thrust of von Baer's embryological observations and the six scholia of development, which von Baer presented in the second half of the first volume of his *Entwickelungsgeschichte*. Von Baer was explicit on this score when he summarized his "most general result" with the aphorism "The development of the individual is the history of increasing individuality in every respect."[29] This notion corresponded with the economic analogy of the division of labor brought into biology by Henri Milne-Edward during the same decade. The twin notions of increasing

---

[28] Churchill, "From Machine-Theory to Entelechy" and "Chabry, Roux."

[29] [Karl Ernst von Baer], *Über Entwickelungsgeschichte der Thiere. Beobachtung und Reflexion* (Königsberg: Bornträger, 1828), pt. 1, p. 265. Facsimile edition (Brussels: Culture et Civilization, 1967).

individuality and the division of labor operated in a variety of biological domains. Darwin applied them in both a phylogenetic and ecological sense, and many post-Darwinian morphologists used them in an embryological sphere. In all senses the two mutually reinforcing notions adduced a dendriform image for the separation or divergence and differentiation or specialization of anatomical parts or species. Species divergence and anatomical specialization became superimposed on the geometry of successive cell divisions so that the different processes merged into a single pattern. The empirical evidence for the first reinforced the mechanical and seemingly automatic process of the second. Cell division, germ layer separation, histological differentiation, and organ specialization and speciation were all manifestations of the same irreversible processes of increasing individuality and of the principle of the division of labor. In the 1870s, Karl Gegenbaur and Otto Bütschli employed both notions in their discussions of development and the evolution of sexuality; in the next decade, Roux and Weismann used them to justify qualitatively unequal nuclear division, but the images remained formal and static.[30] Hertwig wished to employ them in a more functional and dynamic sense. For him the increasing individuality or division of labor did not imply a loss of potentiality; development brought with it a growing network of connections and interactions as much as a geometrical separation among the parts. But the power of Hertwig's insight, I suspect, was blunted by his polemical style and lack of clarity about the causes of physically observed differentiation.

Where Hertwig failed, however, the American school, led by Morgan, fared better. In 1901, Morgan wrote the most comprehensive work on regeneration, and, as Maienschein has pointed out (Chapter 9 in the present volume), he adopted a much more functional view of differentiation. In the same year, Hans Spemann began his own

---

[30] Carl Gegenbaur, *Grundriss der vergleichenden Anatomie* (Leipzig: Wilhelm Engelmann, 1874), p. 90; Otto Bütschli employed the metaphor of the division of labor in order to explain the evolution of the two sexes. This perspective became useful to Weismann as he developed the concept of the isolated germinal track, and then, by 1891, it became part of his explanation of normal development and regeneration. See Bütschli, *Studien über die ersten Entwicklungsvorgänge der Eizelle, die Zelltheilung und die Conjugation der Infusorien* (Frankfurt am Main: Christian Winter, 1876), p. 219; Wilhelm Roux, *Über die Bedeuterung Kerntheilungsfiguren. Eine hypothetische Erorterung* (Leipzig: Wilhelm Engelmann, 1883), and August Weismann "The Duration of Life," in *Essays upon Heredity and Kindred Biological Problems*, 2 vols., ed. Edward B. Pulton, Selmar Schonland, and Arthur E. Shipley (Oxford: Clarendon Press, 1891), 1:1–35. I have commented previously on this relationship in "Guts of the Matter," pp. 211–12.

career in experimental embryology. His early blastomere constriction and lens induction experiments drew directly on the problems and techniques of the previous fifteen years of regeneration studies. Somewhat later, Charles Manning Child used regeneration as a means for examining the effects of isolated factors in development. By some sleight of hand that I do not yet understand, many of the traditional questions of regeneration appear, after the turn of the century, to become preempted by questions about normal development generated by experiments involving transplantations and single-factor analysis. Barfurth continued his reporting on regeneration studies until 1916, but the emphasis of this work had clearly shifted.[31]

[31] For an excellent account of Spemann's early years of experimentation, see Hamburger, *Heritage.*

9

T. H. Morgan's regeneration, epigenesis, and (w)holism

JANE MAIENSCHEIN

Those who know Thomas Hunt Morgan primarily as a well-established geneticist, and those who value the successes in twentieth-century reductionist programs in genetics, will be surprised to learn that Morgan began his career as a wholist.[1] In fact, Morgan clearly did deny the efficacy of reductionism. Indeed, his first two decades of research exhibit a concern with the organism as a whole that has come in more recent decades to be associated with fuzzy thinking and sloppy vitalism. Textbooks often ridicule such biologists gone wrong as Hans Driesch, who finally turned to philosophy rather than biology to support his vitalistic and wholistic position. Modern biology tries to convert such biologists as Morgan, Charles Manning Child, J. S. Haldane, and Jacques Loeb into something other than they were by omitting discussion of their wholistic inclinations.

It is worthwhile to explore such wholistic ideas and to attempt to understand why in the early part of this century they seemed both consistent with good science and necessary. This, in turn, should help to illuminate the perpetual discussion of whether biology should be in some interesting way considered a unique science or whether it is just

---

[1] Key terms such as "wholism," "holism," and "materialism," will be defined in the course of the chapter. These terms have been given a widely varied range of meanings by different writers, which has created and continues to create a great deal of confusion. Garland Allen, for example, insists that Morgan was a "dialectical materialist." For Allen, such materialists are "concerned with the complex interactions between parts in a whole. They see the whole as equal to more than the sum of its parts (i.e., to the sum of the individual parts plus their interactions). They reject the simplistic billiard-ball model of physical processes. They see all processes in the world undergoing constant change, motivated by the interaction of various contradictory (dialectical) elements within the systems themselves." Garland E. Allen, *Thomas Hunt Morgan: The Man and His Science* (Princeton: Princeton University Press, 1978), p. 327. I do not agree that Morgan fits this description, and I offer an alternative interpretation, using the term "wholism" to emphasize the distinction.

133

like physics and chemistry. Focusing on Morgan, who should prove noncontroversially respectable to all parties, rather than on a "suspect" individual like Driesch or a less well-known figure like Child, should demonstrate the pervasiveness of wholistic thinking.[2] It should also reveal the central role that regeneration played in establishing respectability, however fleeting, for that position.

Morgan on development

During the 1890s, Morgan explored a wide range of questions about embryological development of a variety of organisms. Among other projects, he began a survey of studies of frog development, which had become a popular subject in recent years. Europeans in particular had been centrifuging frogs' eggs, putting them under pressure, rotating them within gravitational fields, killing blastomeres, and otherwise subjecting them to various sorts of manipulations. The result was a plethora of data and interpretations. All assumed that the frog's egg is essentially a material thing, with no special vital forces involved in directing its production or development. None of the leading researchers invoked special vital forces or entities to account for embryonic action. All assumed that some combination of motions and forces, either internal or external to the egg, direct development. But not all agreed as to the relative importance of the various possible forces. Nor did all agree that development could be explained simply by reducing the developmental phenomena to the actions of matter in motion. Morgan examined the subject and surveyed the literature in his book *The Development of the Frog's Egg*.[3] Not surprisingly, the project took him longer than he had expected and carried him into a consideration of larger questions about the nature of development and to a level of detail that he had not anticipated. He went to Europe in 1894–5, in part to complete the book.

Inspired by Jacques Loeb, his colleague at Bryn Mawr College in 1891–2, and by his friends from the Johns Hopkins University graduate school and the Woods Hole Marine Biological Laboratory, including Edmund Beecher Wilson, Morgan also began by 1893 to carry out an experimental study of the development of marine invertebrates. He read about the experimental studies of development emerging from laboratories in Germany (which Wilson had just visited for a year

[2] Sharon Kingsland, on Child and Herrick, in progress.
[3] Thomas Hunt Morgan, *The Development of the Frog's Egg: An Introduction to Experimental Embryology* (New York: Macmillan, 1897).

and from which Loeb had come to the United States), and he began to explore the same sorts of questions using the same kinds of techniques. His own visit to the Naples Zoological Station in 1894–5 reinforced his interest and directed him also toward new techniques and questions. There he became good friends with Hans Driesch, who had a provocative point of view and a high level of energy that promoted eager discussion among the young embryologists gathered in Naples.

By 1894, when Morgan met him, Driesch had moved from the strict mechanistic and reductionistic position he had originally held. He had begun to suggest that although an analytic, or mechanistic, account of development was appropriate, perhaps that account need not be reductionistic.[4] Such a distinction calls for clarification of terms.

In the context of this discussion, materialism and mechanism are taken to be primarily ontological positions. They assert that the nature of what exists in the world is material or a combination of matter and motion. Materialists hold that there is no special vital something that exists in addition to matter and motion. There are no vital forces or vital entities in the world. Yet a materialist need not be a reductionist. Instead the materialist might adopt the wholist position that what exists in the world is complex, interactive wholes; alternatively the materialist might insist that what exists is really parts, which have reality of their own. For "reductionism" is taken here as an epistemological position, referring to the reduction of theories, explanations, or sentences.[5] The reductionist biologist claims that all of life's phenomena can be reduced to – that is, explained in terms of – physics and chemistry. Yet others might hold that there is something about living phenomena that goes beyond the basic properties and laws of physics and chemistry. In particular, the whole might be more than the sum of the parts; that is, there might be something about the nature of the whole organism that makes it act in a way that cannot be explained in terms of the sum of the actions of the parts. This view is generally labeled "holism." Driesch was beginning to explore this holistic position by 1894, as well as a wholistic ontology. So was Morgan.

---

[4] For example, Hans Driesch, *Analytische Theorie der organischen Entwicklung* (Leipzig: Wilhelm Engelmann, 1894).

[5] There is a vast literature on mechanism, vitalism, and reductionism, much of it quite unilluminating and confusing. In large part this stems from the varied and contradictory use of the terms by different writers, or even by the same writers. I use the terms in a way that seems to make sense, coincides with their meanings in embryology around 1900, and remains consistent with the major historical and philosophical discussion of the subject.

Morgan's move to regeneration

By 1895, Morgan realized that it was difficult to get at the basic processes of normal development simply by peering at developing embryos. Yet radical, highly interventionist experimentation raised the question of whether the resulting process remained sufficiently like normal development to provide any useful information. Regeneration experiments, however, seemed to mimic natural conditions and could therefore provide information about normal developmental processes.

In a paper of 1896, Morgan began thinking about regeneration. He first reported on his own and other studies of partial embryos.[6] In 1888, Wilhelm Roux had stimulated a succession of explorations of the frog's egg in which one of the first two, or four, or eight blastomeres was pricked with a hot needle and thereby, evidently, killed.[7] The resulting partial embryo could still develop, Roux showed, but only into the part that it normally would have become. The part could not compensate for the absence of the rest of the material. Or so Roux had believed at first. Yet time produced conflicting results, and Roux later concluded that the part is capable of a process of "postgeneration" that resembles the normal regeneration process and by which the part "re"-generates the whole.

In contrast, Driesch's study of sea urchins, initially intended to support and extend Roux's results for frogs, had shown clearly by 1892 that partial embryos can indeed develop into whole larvae.[8] The part is capable of producing the whole and actually does so. This suggested that something about the nature of the whole was available in the part, or at least that the wholeness could be regenerated after injury. But how? And was a certain minimal size of material or minimal number of cells required before regeneration could occur? Driesch, Wilson, Morgan, and a handful of others took on these questions in the 1890s.

---

[6] Thomas Hunt Morgan, "Studies of the 'Partial' Larvae of *Sphaerechinus*," *Archiv für Entwickelungsmechanik der Organismen* (1895) 2: 81–126.

[7] Wilhelm Roux, "Beiträge zur Entwickelungsmechanik des Embryo. Über die künstliche Hervorbringung halber Embryonen durch Zerstörung einer der beiden ersten Furchungskugeln, sowie über die Nachentwickelung (Postgeneration) der fehlenden Körperhälfte," *Virchow's Archiv* (1888) 114: 113–53; partly translated in B. Willier and J. Oppenheimer, eds., *Foundations of Experimental Embryology* (Englewood Cliffs, N.J.: Prentice-Hall, 1964), pp. 2–37.

[8] Hans Driesch, "Entwicklungsmechanische Studien. I, Der Werth der beiden ersten Furchungszellen in der Echinodermentwicklung. Experimentelle Erzeugen von Theil- un Doppelbildung," *Zeitschrift für wissenschaftliche Zoologie* (1892) 53: 160–78 and 183–4; translated in Willier and Oppenheimer, *Foundations*, pp. 38–50.

Morgan concluded, after carefully surveying the literature and his own accumulating results, that partial larvae of sea urchins behave like the material in normal and familiar cases of regeneration. The whole is regenerated from the part, just as surely as a worm can regenerate its tail or a crab its leg. Furthermore, it must be conditions internal rather than external to the organism itself that drive development, despite the pressures from outside the egg that have caused the artificial condition.

But the process of regeneration is not just a simple mechanical or chemical playing out of preexisting directions, Morgan concluded. Rather:

In much the same way, an animal or plant tends in many cases to replace a part of itself that has been lost or injured by external agencies; i.e. we say the whole is regenerated from a part. We can find no chemical or physical explanation for any of these phenomena. It does not make our problem any easier to admit the possibility that factors may be present in the ontogeny that are dependent on principles unknown and unrecognized by the chemist and physicist. We call these "vital" factors and in many of the fundamental problems of Biology, such as development, cell-division, and regeneration these vital processes come to the front. So far as we can see at present the vital factors that control the development do make use of many known chemical and physical properties of matter, but it seems to me that it is very rash at present to conclude therefore that the vital processes of living things are necessarily only the complex of known physical and chemical processes.[9]

There was no legitimate reason to assume that special vital somethings exist, but the living phenomena might be the product of something more than the sum of known physical and chemical processes. Morgan certainly left the door open to vitalism, although he inclined more to a materialistic wholism. As yet, however, he sought to remain agnostic. He simply did not think there was sufficient scientific evidence to permit researchers to determine either the ontological or the epistemological question definitively one way or the other.

By 1897, Morgan had read August Weismann's theoretical explanation of regeneration as a process derived from evolution by natural selection, had decided that it was entirely unsupported, and had turned directly to investigating regenerative phenomena himself.[10]

[9] Morgan, "The Number of Cells in Larvae from Isolated Blastomeres of *Amphioxus*," *Archiv für Entwickelungsmechanik der Organismen* (1896) 3: 269–94 (quoted from p. 292).
[10] Morgan, "Regeneration in *Allolobophora foetida*," *Archiv für Entwickelungsmechanik der Organismen* (1897) 5: 570–86. Weismann then responded to Morgan in a paper widely available to his English-speaking audience: "Regeneration: Facts and Interpretations," *Natural Science* (1899) 14: 305–28.

Following up studies by Loeb and Driesch on planarians, Morgan discovered the animals' ability to regenerate the normal parts under a wide variety of complicated conditions. This showed that they did not just add on new material after an injury in order to replace the missing old material. Rather, they actually had the power to transform the old material into a new part. "We see here not only a power of regeneration," Morgan concluded, "but also a subsequent self-regulation, and by means of the latter the normal relations of the parts, characteristic for the species, are regained." Surprisingly, the 'body material of the already formed planarian remains "almost as plastic as that of an undivided or dividing egg."[11] Something about the nature of the whole made it possible for the individual to transform material and regulate the characteristics of the whole. Morgan called this process "morphallaxis."

In another paper of 1898, Morgan took on Weismann's theory directly. Weismann had argued that certain parts of an organism are more susceptible to injury than others. These, influenced over the course of time and by the constant action of evolutionary forces, gain a power to regenerate because of a built-in set of auxiliary inherited materials that guide the reconstruction of the part in question. Thus, according to Weismann, ability to regenerate is correlated with liability to injury. This view was particularly appealing because it implied that an injured organism only needed to kick into action a preexisting and inherited capacity for regenerating a missing part. It did not have to go through a complex epigenetic process of determining what part was needed and then generating it anew. Yet this predeterministic interpretation did not appeal to a committed epigenesist like Morgan.

In a series of studies, Morgan sought to establish that this view is false and that regeneration is akin to the normal developmental processes, rather than a special ability adapted for special cases. By showing in hermit crabs, for example, that the parts most likely to be injured are not those most able to regenerate, he undercut Weismann's theory.[12] If Weismann was wrong, however, the question remained, What does cause regeneration? And this is really the key. For Weismann's preformationist account could easily explain regeneration if the injured organism simply grows the right sort of part by adding new material as directed by hereditary information. Morgan

---

[11] Morgan, "Experimental Studies of the Regeneration of *Planaria maculata*," *Archiv für Entwickelungsmechanik der Organismen* (1898) 7: 364–97 (quoted from pp. 389, 396).

[12] Morgan, "Regeneration and the Liability to Injury," *Zoological Bulletin* (1898) 1: 287–300.

did not accept that idea, and he therefore had to provide a much more difficult, epigenetic account of regeneration, as well as of generation in the first place.

Morgan spent part of the summer of 1898 in Europe, visiting Naples and elsewhere. On his return trip to the United States in September, he found himself with time on his hands in London when his ship was delayed. He went to the British Museum and read classic eighteenth-century regeneration studies, particularly those of Trembley and Bonnet. As he wrote to Driesch, he found these works much more interesting than contemporary work on the subject.[13] These eighteenth-century theories became the starting point for Morgan's systematic treatment of regeneration.

In lectures he gave at the Woods Hole Marine Biological Laboratory in Massachusetts in 1898 and 1899, Morgan considered the various alternative theories and the available data about regeneration. The fact that a part can regenerate a new whole in a way parallel to the action of normal development raised a most interesting problem, namely to explain how that epigenetic response to abnormal conditions is possible. Morgan concluded that

something more is included in these phenomena . . . than can be explained by simple physical interaction or by chemical influences. The process that takes place suggests that something like an intelligent process must be at work – I mean that what we call correlation of the parts seems here to belong rather to the category of phenomena that we call intelligent, than to physical and chemical processes as known in the physical sciences. The action seems, however, to be intelligent only so far as concerns the internal relations of the part, i.e., it acts rather as a "perfecting principle" than as a process of adaptation to external needs (adaptation).[14]

Morgan realized that to account for this "perfecting principle" in a way that was not vitalistic but that provided a real – that is, a legitimate – scientific explanation was the central problem for the study of embryology and regeneration.

Furthermore, not only might an extraphysical, intelligent process be at work, but normal reductionist methods from the physical sciences might prove inadequate to the biological task. Morgan explained:

[13] Morgan to Driesch, September 13, 1898, cited in Allen, *Morgan*, pp. 86–7.
[14] Morgan, "Some Problems of Regeneration," *Biological Lectures Delivered at the Marine Biological Laboratory of Woods Hole* (hereafter cited as *Biological Lectures*), (1899) 1898: 193–207 (quoted from pp. 205–6).

In much of our biological work we have been guided by methods derived from the physical sciences, and most fortunately so, for perhaps only in this way can we hope to reduce living phenomena to simpler terms. But sooner or later we meet with a factor that defies further physical analysis, and this factor seems to be present in all biological phenomena. We gain nothing by calling it a vital force, unless we can define what we mean by vitality. Whether or not this factor* [*Morgan's footnote:* "*It is simpler to speak of it as one factor, but it may equally well be true that there are many factors."] is only a complex of physical forces that we cannot unravel, or whether there exists something that cannot be expressed in terms of physics and chemistry – that is the question!

We err, I think, in going at present to either extreme, i.e., either in ignoring this something that has been called a vital force and pretending that physics and chemistry will soon make everything clear, or, on the other hand, in calling the unknown a vital force and pretending to explain results as the outcome of its action.

In our studies of the development of form we meet most often with this factor. Are we at bottom trying to give a causal explanation of form itself, and, if so, is not our problem insoluble? Can we hope to do more than determine under what internal and external conditions a given form appears? If we limit our researches to this problem we can hope to succeed. But can we go back of this and explain the reaction itself? At present we have not succeeded in doing so, any more than has the mineralogist explained the form of the crystal. It may be that what we call a formative force or a vital force is the property of living things to assume a given form under certain conditions. If so, is there here legitimate ground for investigation, or rather let me ask, can we hope to extend our investigations beyond the knowledge of the internal and external conditions within which new forms arise. It is this uncertainty in regard to the problem of vitality that we need first to clear up, and it seems to me that this is the cardinal point for us to examine at present. It is possible, I think, by means of experiment alone, to determine how far and in what sense we can pursue the investigation of the causes of form. In this regard experimental studies on the regeneration of animals and plants offer a most admirable field for future work.[15]

And so Morgan launched into an experimental study of regeneration, seeking a "verifiable hypothesis," since he felt that only in this way could progress be made in science.[16]

Experiments with hydromedusae, earthworms, more with planarians, studies of frogs (including grafting of pieces between two different species, to determine which tissue gives rise to the regenerated part) – all contributed to Morgan's suggestions in his 1899 Woods Hole lecture. There he discussed earlier theories of regeneration, considered the alternatives in more detail than during the preceding summer, addressed current work on the subject, and formulated what

---

[15] Ibid., pp. 206–7.
[16] Morgan, "Regeneration in *Hydromedusa, Gonionemus vertens*," *American Naturalist* (1899) 33: 939–51, p. 950, calls for such a hypothesis.

he described as a tentative and temporary working hypothesis. What we call "regeneration" involves many different phenomena, he had concluded, and this makes any theory of regeneration difficult to achieve. In particular, the phenomena of "morphallaxis," as Morgan called it, or the remodeling of existing material into new parts, caused problems for a simple interpretation. It would be far easier to explain how the organism might add on new material of the proper sort than to account for how already existing and already differentiated material parts become transformed in some coordinated way that retains the organism as a whole.

The central phenomenon, Morgan suggested, was the way in which an injury, however it is caused, stimulates molecular changes within the whole body of the injured organism. "It is this molecular change that, dominating the subsequent development, seems to control it, and gives us the impression of formative processes at work." This might sound mystical, he realized, and yet "the formative processes are only the expression of the physical, molecular structure that has been assumed by the piece."[17] Although a full theory remained to be developed, he felt that the molecular (and hence physicochemical) nature and the organization of the whole seemed crucial to explaining development. He had made progress toward a mechanical account since the previous talk of a vague "perfecting principle," but he was no closer to a reduction of the explanation to physical and chemical terms.

The next year brought further results and further suggestions. Perhaps the physical or chemical structure of the original part is carried directly over into the new material that is bringing about the regeneration; perhaps there are formative "stuffs" or "actions" of some sort. But Morgan rejected this view because the material retains its plasticity and therefore cannot have been so completely specified that the specification alone could explain its subsequent development into a particular part. The new material remains too "omnipotent" for such an account to work. Vague as the conclusion necessarily remained, Morgan felt that he must "suppose that there is something in the structure or composition of the head and tail, that may act as a determining factor in the production of heteromorphic regeneration."[18] The organization was key.

Yet the nature of this structure or organization remained the critical point, and Morgan continued to attack that problem. In a more gen-

---

[17] Morgan, "Regeneration: Old and New Interpretations," *Biological Lectures* (1900) 1899: 185–208 (quoted from p. 207).
[18] Morgan, "Regeneration in Planarians," *Archiv für Entwickelungsmechanik der Organismen* (1900) 10: 58–119 (quoted from p. 113).

eral article of 1901, he compared the phenomena of regeneration in the egg, the embryo, and the adult. All exhibit parallel processes, he concluded, and all show that some organization of the whole organism lies behind and directs development.

Somehow a piece of an egg, for example, which can have no inherited or preestablished bilateral symmetry, since it is only part of a normal whole, nonetheless manages to give rise to a properly bilateral embryo. Since it clearly cannot have relied on a preestablished symmetry, it must have assumed that symmetry. Regeneration must have been epigenetic, rather than preformed, and cannot have resulted from any simple inherited structure or stuffs. Yet this "gives us one of the most interesting and also important problems with which the student of experimental embryology has to deal. We know of nothing similar taking place in inorganic nature."[19]

Crystal formation is not parallel, since the crystal never has to reorient and rearrange itself in this way. Nor is the way in which pieces of a magnet reorient themselves analogous, for, Morgan continued,

our conception of the polarity of the magnet rests on the idea that it is the sum total of the polarities, or, perhaps, of the orientation of the minutest elements, the molecules, of which the magnet is made up, while our conception of the organization of the egg is exactly the reverse (or at least I shall try to show that we must really believe this to be the case), and we must think of the entire egg as a whole and not the sum total of an infinite number of smaller wholes. We may claim, I think, that this property of the egg substance of forming itself into a new whole is peculiar to the living protoplasm and is a property that we do not find, or have not found as yet, in inorganic, or perhaps we may go further and say in dead, matter. If we choose to call this property of living matter a vital factor in the sense that it is not found in matter that is dead there can be, I think, little objection to so doing. If the statement seems to be arguing in a circle, we may state more simply that those properties of living things that are not shown by non-living things we shall call vital properties. We may add that we cannot be sure, at present, whether these vital factors will conflict with our present ideas of causality or not; they seem rather to be, however, new causal phenomena peculiar to certain organic substances or compounds, but it would be out of place here to examine further into these difficult questions.[20]

And further:

We are therefore, I believe, also justified in calling the organization of living things a vital property in the sense, to repeat what I have just said, that it is

[19] Morgan, "Regeneration in the Egg, Embryo, and Adult," *American Naturalist* (1901) 35: 949–73 (quoted from p. 956).
[20] Ibid., p. 957.

peculiar to this kind of substance or structure, and not the result of a complex of known physical principles; or, in other words, it is a physical phenomenon as fundamental as the polarity shown by crystals or the magnetism of the magnet, and just as the latter are associated with certain kinds of matter, so is the organization associated with the substance protoplasm.[21]

This emphasis on organization and the nonphysical nature of that organizational property did not mean that Morgan had succumbed to vitalism, however. He remained strongly inclined to a materialist ontology even though he recognized the difficulty of supporting this point of view. He wrote to Driesch, in response to a paper of Driesch's on vitalism:

I had intended reading again your paper on vitalism in order that I might write to you more specifically about it. I do not dare begin, however, for my letter would never come to an end. I follow you a long way, but cannot truthfully say that I consider you have "proven" the existence of a principle of vitality – except insofar as there is much that we cannot explain. Loeb thinks, and I am almost prepared to follow him, the idea is a sterile one in regard to its value as a working hypothesis; at any rate, it comes dangerously near to metaphysics in our present state of knowledge. That, however, is only a matter of opinion.[22]

Regeneration

In 1900, Morgan was invited to deliver a set of lectures at Columbia University. He later expanded those five lectures into a book, *Regeneration,* where he maintained the conclusion that he had been developing throughout his regeneration studies. The factor that allows the injured part of an organism to undergo reorganization and make up a new, functioning organism of the right sort is the internal organization of the whole. The key question, then, must still be "what is organization." Yet: "This it must be admitted is a question that we cannot answer. Looked at in this way the problem of development seems an insoluble riddle; but this may be because we have asked a question that we have no right to expect to be answered."

If the physicist were to ask what gravity is, he would ask just such a question and would likewise have no right to expect an answer. But he can ask other, more accessible questions about gravity, such as what its effects are. So with embryonic development and organization. Yet there is still something about organic organization that seems not

---

[21] Ibid., p. 973.
[22] Morgan to Driesch, September 13, 1899, quoted in Allen, *Morgan,* pp. 321–2.

parallel to the case of gravity, he argues: "The action of the organism is sometimes compared to that of a machine, but we do not know of any machine that has the property of reproducing itself by means of parts thrown off from itself."[23]

The organization of the developing organism is more complex and demands more study. That study must remain the very center of embryology. For:

It can be shown, I think, with some probability that the forming organism is of such a kind that we can better understand its action when we consider it as a whole and not simply as the sum of a vast number of smaller elements. To draw again a rough parallel; just as the properties of sugar are peculiar to the molecule and cannot be accounted for as the sum total of the properties of the atoms of carbon, hydrogen, and oxygen of which the molecule is made up, so the properties of the organism are connected with its whole organization and are not simply those of its individual cells, or lower units.[24]

Morgan realized that getting at just what this organization means, what causes it, and what effects it brings was basic to embryology. He also recognized the difficulty of that central task. He acknowledged toward the end of the regeneration volume that "there must be a certain amount of vagueness connected with our idea of what the organization can be." It is the structure of the organism "to which are to be referred all the fundamental changes in form" but about which we admittedly know little. In fact, "we know this organization at present from only a few attributes that we ascribe to it, and we are not in a position even to picture to ourselves the arrangement that we supposed to exist."[25]

Morgan accepted Du Bois-Reymond's "Ignorabimus," according to which he acknowledged that there are things in science that we do not now know and things that we shall probably never know. Morgan clearly preferred not to think that this organization was one of the unknowables, but he was prepared to accept that conclusion if forced to do so. In the meantime, he proposed to continue seeking experimental evidence from regeneration studies and elsewhere to provide insights into just what this organization is and how it works.

In the course of his next decades of research on embryology, Morgan discussed the process of differentiation in terms of tensions and pressures within the organism. He suggested, as a working hypothesis, that these mechanical forces might account for organization, although he recognized the rather unspecific nature of the hypothesis.

[23] Morgan, *Regeneration* (New York: Macmillan, 1901), pp. 358–9.
[24] Ibid., pp. 278–9.
[25] Ibid., p. 288.

The epigenetic perspective

During the early decades of this century, a number of embryologists – particularly the Americans – joined Morgan in calling for an epigenetic and wholistic view of development. Frank Lillie, for example, urged that embryologists consider the "properties of the whole" that produce a "principle of unity of organization."[26] Scripps Institution of Oceanography director William E. Ritter urged study of the unity of the organism.[27] Charles Otis Whitman insisted many times that the cell could not be considered the fundamental unit of life, if that is taken to mean that understanding the cell will explain all phenomena of life; instead, the interactions of cells are essential and demand an organizing principle.[28]

Edmund Beecher Wilson remained more ambivalent. His specialization in cytology tempted him to view the cell as the basic life unit, yet even Wilson urged that the connections among cells in the organism are crucial and that "the life of the multicellular organism is to be conceived as a whole."[29] He continued: "The only real unity is that of the entire organism, and as long as its cells remain in continuity they are to be regarded, not as morphological individuals, but as specialized centres of action into which the living body resolves itself, and by means of which the physiological division of labor is effected."[30] By 1923, however, Wilson acknowledged that when we invoke the action of the "organism as a whole" or some "principle of organization," in fact we mean that "we do not know."[31]

Jacques Loeb held to a strictly mechanistic interpretation of life and denied the existence of a special "directive force" that explains organization. Yet he concerned himself with the organism as a whole and sought to account for the interactions of the various parts in terms of chemical reactions and interactions. Like Morgan, Loeb saw regeneration as a central problem for embryology. Asking what causes injured parts to regenerate only when injured and not otherwise – in other words, asking how the part knows to regenerate – Loeb concluded

[26] F. R. Lillie, "Observations and Experiments Concerning the Elementary Phenomena of Embryonic Development in Chaetopterus," *Journal of Experimental Zoology* (1906) 3: 153–268, especially the section "Properties of the Whole."

[27] William E. Ritter, *The Unity of the Organism* (Boston: Badger, 1919).

[28] Charles Otis Whitman, "The Inadequacy of the Cell Theory of Development," *Journal of Morphology* (1893) 8: 639–58.

[29] E. B. Wilson, *The Cell in Development and Inheritance* (New York: Macmillan, 1896), p. 41.

[30] E. B. Wilson, "The Mosaic Theory of Development," *Biological Lectures* (1894) 1893: 1–14 (quoted from p. 9).

[31] E. B. Wilson, *The Physical Basis of Life* (New Haven: Yale University Press, 1923), pp. 45–6.

that "certain substances" were involved. These "certain substances" circulate throughout the organism. They accumulate and prevent further growth when the organism is completely developed. Then injury can cause them to flow into the hurt part and to stimulate regrowth or regeneration. Thus, according to Loeb, it is "utterly unnecessary to endow such organisms with a 'directing force' which has to elaborate the isolated parts into a whole."[32] The movement of chemical substances can explain all.

Charles Manning Child developed a different point of view, but one responding to the same impulses. Individuality of the whole organism is achieved through the establishment of a system of gradients in the individual. The gradients develop, become more strongly established, and eventually gain virtual permanency. At that point, in conjunction with other, related changes, the parts of the organism have been made into an interacting whole. "From this point of view," Child concluded, "the assumption of a mysterious, self-determined organization in the protoplasm, the cell or the cell mass as the basis of physiological individuality becomes entirely unnecessary."[33] For Child, the internal structure determined by the set of gradients and the responses to the external world can explain normal development and regeneration alike.

A host of others made similar calls for an epigenetic interpretation of development, in which no appeal was to be made to the preexistence of inherited units of information that simply unfold or grow up to bring into existence the adult differentiated being. This emphasis was accompanied by other demands to understand epigenetic development in terms of the whole organism, rather than simply as the sum of the individual parts. The fundamental nature of life seemed, to many biologists in the early decades of this century, to require such epigenetic and wholistic thinking.

The next decades brought new lines of research trying to get at the same phenomenon of organization. Carl Vogt's use of advanced vital staining techniques established a detailed fate map for the frog's egg, for example. This showed that normal eggs are already highly specified, so that materials from different locations in the egg become particular germ layers and particular parts of the embryo and adult with great regularity. Normal development is highly directed by the structure and materials of the egg. This made the phenomena of

[32] Jacques Loeb, *The Organism as a Whole from a Physicochemical Viewpoint* (New York: Putnam, 1916), pp. 9–10.
[33] Charles Manning Child, *Individuality in Organisms* (Chicago: University of Chicago Press, 1915), p. 41.

regeneration all the more puzzling and all the more important, for discovering how the egg regenerates in response to disruptions might well reveal the developmental processes by which it became so highly directed in the first place.

Hans Spemann's work on tissue transplantation also held significant promise for explaining both normal development and regeneration. By 1918, Spemann had begun adapting Gustav Born's technique of transplanting pieces of embryonic tissue from one individual to another. With differently pigmented members of different species, the experimental technique allowed the observer to watch the relative contributions to the resulting embryo from each of the pieces of tissue. For example, suppose one takes a piece of tissue that would normally give rise to limb, removes it from the host, and grafts it onto a donor. Does the developmental process that occurs involve tissue from the host, the donor, or both – and how so? The fact that tissue from both host and donor contribute and yet the result is a normally structured and functioning individual shows that the development is not simply predetermined. A fairly sophisticated regulation of the whole must occur.

Spemann also experimented with eggs and with early embryonic tissue. The results of his and others' work showed that the dorsal lip of the blastopore was especially "potent." In 1921, Spemann suggested to his student Hilde Mangold that she should try transplanting pieces of the dorsal lip of the blastopore into a host gastrula. The resulting induction of a secondary embryo at that point suggested that the development worked through a kind of chemical action, not unlike Child's gradients. Spemann developed the concept of the "organizer," which "creates an organization field of a certain [axial] orientation and extent, in the indifferent material in which it is normally located or to which it is transplanted."[34] Spemann certainly emphasized organization of the whole as a responsive and interactive unit, yet he avoided vitalism.[35] He also held an epigenetic view of development.

---

[34] Hans Spemann, "Die erzeugung tierischer Chimären durch heteroplastische Transplantation zwischen *Triton cristatus* und *taeniatus*," *Roux's Archiv für Entwickelungsmechanik der Organismen* (1921) 48: 533–70 (quoted from p. 568); translated in Viktor Hamburger, *The Heritage of Experimental Embryology: Hans Spemann and the Organizer* (New York: Oxford University Press, 1988), p. 45.

[35] Hamburger, *Heritage*, pp. 65–7; T. J. Horder and P. J. Weindling, "Hans Spemann and the Organiser," in T. J. Horder, J. A. Witkowski, and C. C. Wylie, eds., *A History of Embryology* (Cambridge: Cambridge University Press, 1985), pp. 183–242.

Eventually the concept of the organizer proved problematic, for it was found that a wide variety of substances could induce embryonic change. After the apparent triumph of genetics in the 1950s, embryologists tended to view development in terms of genetics, stressing the action of genetic messages, through developmental genetics. Few researchers have persisted in searching for explanations of how the whole organism maintains its individuality and its ability to regulate and regenerate. Reductionism has come to appear more promising.

As Wilson said, talking about organization amounts to admitting that "we do not know," but scientists always desire to advance knowing rather than deal with not knowing. Increasingly, science has come to be judged in terms of its products – bits of knowledge yielding definite and practical results. Morgan himself accepted this as the goal for science and set off on his productive research program in genetics, which did yield answers.

Yet Morgan remained committed to studying development in terms of epigenesis and of at least whole regions of the organism, if not the whole organism. In his *Embryology and Genetics* (1934), for example, he considered regeneration and asked how we can explain how cells with identical genetic makeup undergo regeneration and reorganization into the appropriate sort of new tissue and thereby allow regeneration into a full organism. He concluded: "The old act as the determiners or organizers of the new, and we can perhaps assume that this is a chemical influence. It is not safe, however, to push such comparisons too far." Induction or action of organizers or determinants was apparently at work; yet just how was not clear. He admits: "The nature of these prospective influences is entirely unknown at present, and may be very different in kind and degree in different parts of the embryo."[36] By 1934, Morgan was focusing more on regions than whole organisms, but he clearly maintained his emphasis on the whole and on epigenesis and resisted the move toward reductionism in embryology.

Conclusion

Despite the accelerating move away from wholism and toward reductionism in embryology, and in biology in general, questions remain, and the issue has not yet been settled for all time. We still cannot fully account for the remarkable ability of the organism to regulate and

---

[36] Morgan, *Embryology and Genetics* (New York: Columbia University Press, 1934), pp. 169, 198.

regenerate during development. We may eventually find an explanation in terms of the chemical makeup and physical processes in the organism. Certainly studies on cell–cell interactions, for example, hold promise. Whether we can in fact achieve this reduction remains an open question, and whether we can do so in principle remains a matter of conviction rather than rigorous logical demonstration.

There was something about the ability of the organism to regenerate that led most American embryologists around 1900 to believe in an organization of the whole that amounted to more than the sum of the parts. There may also be something about the nature of our nervous system that, as gestalt psychologists suggest, inclines us to think synthetically rather than analytically and therefore to see the world more easily in terms of wholes than in terms of parts. At any rate, the current bias toward reductionism has not settled the issue, any more than has the morass of writing on reductionism by philosophers of science, especially during the 1960s and 1970s.

Ernst Mayr argues that biology is an autonomous science, in part because living things act as wholes ("as if they were a homogeneous entity") rather than as sums of parts; "their characteristics cannot be deduced (even in theory) from the most complete knowledge of the components, new characteristics of the whole emerge that could not have been predicted from a knowledge of the constituents."[37] Morgan would have neither agreed nor disagreed absolutely, but he would have been sympathetic, as would generations of embryologists. Clearly there is still room for epigenesis and wholism in biology, and it may well be phenomena of regulation in regeneration that best show how they fit.

ACKNOWLEDGMENTS

Thanks to Charles Dinsmore and Richard Creath for comments and suggestions on an early draft. This work results from research supported in part by the National Science Foundation, grant SES-87-22231.

[37] Ernst Mayr, *Towards a New Philosophy of Biology* (Cambridge, Mass.: Harvard University Press, 1988), p. 15. For responses, see, for example, Alexander Rosenberg, *The Structure of Biological Science* (Cambridge: Cambridge University Press, 1985), pp. 21–5.

10

A history of bioelectricity in development and regeneration

JOSEPH W. VANABLE, JR.

The origins of the realization that electricity has an effect on the development and regeneration of living tissue are mingled with the earliest awareness that electricity exists and with the beginnings of an understanding of the nature of electricity. Well before the advent of galvanometers, voltmeters, and oscilloscopes, the existence of electricity was detected by observing its effects on living tissue. These effects, seen in both dramatic and subtle ways by so many for thousands of years, have made electricity, in one form or another, an all-pervasive aspect of our lives and of those of legions before us. It has been natural, therefore, for biologists to wonder whether electricity could affect development. In this essay, I propose to trace the notion that electricity might play a role in development and regeneration from its earliest beginnings in the often dubious use of electricity for medical purposes, through the emergence of an understanding of the electrical basis of nerve and muscle physiology, which prepared biologists to imagine a developmental role for electricity, to present-day work examining this issue. (For another account of the early uses of electricity in medicine and physiology, see Borgens, 1989b.)

Knowledge of electrical phenomena in ancient times

As Kellaway (1946) has so beautifully documented, the ancient Egyptians probably knew the effects of electricity, although they almost certainly did not understand the nature of these effects and apparently did not write anything about them. The tombs of their buried leaders, however, were decorated with pictures of the Nile catfish, *Malopterurus electricus,* one of the several electric fish that, as we shall see, appear in the writings of later ancients (Figure 10.1).

The electric catfish appears in the writings of Hippocrates around 400 B.C. (Kellaway, p. 117). He knew it as the *narke* (the Greek root for "narcosis"), no doubt in recognition of the numbness with which

151

Figure 10.1. Nile fishermen, perhaps about to experience a jolt from the specimen of *Malopterurus electricus* (*center*, with barbels and paddle-shaped tailfin) that is in their catch. (Plate 2 in Kellaway, 1946; original kindly furnished by Peter Kellaway.)

such a fish can afflict anyone it touches. Oddly, though, Hippocrates focuses on the nutritive value of this fish rather than its shock value. These fish are sluggish; no need for fright and flight for them, as Cicero realized three hundred years later in pointing out the built-in biological defense mechanism they possess (Kellaway, p. 120); their muscles, Hippocrates says, are quite tender and easy to digest. So according to Hippocrates, for asthma take one serving of catfish each morning, well cooked, with no danger of shock. Hippocrates tended to avoid the "marvelous and esoteric" (pp. 124–7).

As was implied by the name they gave this fish, however, the Greeks of the fourth century B.C. and later were well aware of the jolt an electric fish could deliver to a person. In Plato's dialogues, Meno roguishly accuses Socrates of "torpifying" him as might a torpedo fish (another common electric fish), and Aristotle notes the narcotizing power of the torpedo in his *Historia animalium.* Theophrastus, a disciple of Aristotle's, recognizes that the shock from the torpedo could be conducted through the trident used to spear it. Similarly, Plutarch records the fact that this effect can be transmitted through the water itself. Pliny the elder, somewhat later, speaks of the feet of even the best runner being riveted by the shock delivered through the spear that has hit a torpedo (pp. 118–20).

Early medical uses of electricity

By the first century A.D., the earliest records of the medical use of the electric fish's electricity appear in the writings of Scribbonius Largus, who recommends shock treatment for headache and gout. Kellaway notes this as the first departure from the conservative Hippocratic approach to medicine, and a backward step at that. But the floodgates were open, and by about A.D. 70, the Herbal of Dioscorides adds headache and prolapsed anus to the list of maladies treatable by jolts from the torpedo (Kellaway, p. 130).

By the end of the second century, or early in the third, Galen began to ponder the cause of the numbness brought about by creatures like the torpedo fish. He speculated, tautologically, that it was the "frigorific" principle of the animal spirit coursing through the nerves of these creatures that was responsible (Kellaway, p. 123). Galen, however, was better as a skeptical empiricist than theoretician. He doubted that shock therapy could be effective as a remedy for headache, gout, and prolapsed anus. He therefore tested the claim for headache on himself and pronounced the treatment a success.

Development of techniques for generation, storage, and delivery of electricity

All this while, another form of electricity – static electricity – was becoming known. In the fifth century B.C. Thales, one of the Seven

Wise Men of Greece, recorded the fact that if amber (or *elektron,* as the Greeks called it) is rubbed with cat's fur it produces the amber – or, in Greek, the "electric" – effect (Geddes, 1984): The fur crackles, and the amber picks up lint, bits of parchment, and the like. Quite some time later, early in the second half of the seventeenth century, Otto Von Guericke produced the first machine to generate static electricity, a rotating sulfur ball against which the operator's hand was held to effect the transfer of electrons (Stainbrook, 1948, p. 157; Geddes, p. S-3). The later improvement of static electricity generators, culminating in the one produced by Ramsden in 1768, along with the invention of the capacitor to store the charge generated by them, set the stage for serious study of the nature of electricity and, for better or for worse, its medical, physiological, and developmental effects on organisms (Geddes, p. S-3).

The capacitor, or Leyden jar, as it came to be called, was probably invented by Ewald von Kleist, dean of the cathedral at Carmin in Pomerania, shortly before it was independently discovered in Leyden, in 1745, by Pieter van Musschenbroeck (Hoff, 1936, p. 161; Stainbrook, 1948, p. 157). Much higher voltages could be developed in these devices, and therefore much stronger shocks could be delivered from them than had been possible with the static electricity generators commonly used to charge them. L'Abbé Jean-Antoine Nollet liked to demonstrate this in dramatic fashion (Hoff, p. 161). On one occasion at Versailles, he caused one hundred eighty of the king's guards to leap simultaneously by having them all hold hands and then connecting the man on one end of the line to a charged Leyden jar. This novelty was greatly amplified by conducting a similar exercise with the whole population of a Carthusian monastery, which is said to have strung out a mile's worth of humanity that leaped in concert on receiving the charge contained in a Leyden jar.

Resurgence of electrotherapeutics and the beginnings of physiology

Electrotherapeutics gained new vigor with the advent of practical electrostatic generators and the Leyden jar. In 1744, Johann Krueger, a professor of medicine at the University of Halle, began to use primitive electrotherapy, and by 1755, Richard Lovett claimed to be treating mental disease successfully with electric sparks and current; in 1759, none other than John Wesley, the founder of Methodism, felt justified in lauding Lovett's work: "I doubt not but more nervous disorders would be cured in one year by this single remedy than the whole English Materia Medica will cure by the end of the century" (quoted in Stainbrook, 1948, p. 157). In the United States, Benjamin Franklin was reluctantly yielding to the clamor for help from ailing members of the local citizenry. In a letter to physician John Pringle in 1757, he described his efforts to revive lame limbs with two 6-gallon Leyden jars. These treatments produced results that were at first impressive. He found, however, that after about a week of treatment, the subjects, discouraged by the severity of the shocks and the leveling off of the initial good effect, typically went home and relapsed to their former state, never to return (Geddes, 1984).

Knowledge of basic physiology was also growing during these times. In 1700, Guichard Joseph Duverney carried out the first electrical stimulation of frog muscle. Fifty years later, at Halle, Christian Kratzenstein stimulated human muscle contractions with static electricity discharged from a Leyden jar; Jean Jallabert both exercised normal human muscle and caused muscle in a paralyzed arm to move; and at Bologna, Marc Antonio Caldani, successor to Giovanni Battista Morgagni's chair in anatomy and teacher of Luigi Galvani, conducted experiments involving the electrical stimulation of nerves (Hoff, 1936, pp. 162–3; Licht, 1944, pp. 451–2; Stainbrook, 1948, pp. 157–8). Between 1750 and 1780, no fewer than twenty-six articles or reviews of books on medical electricity had appeared, and by 1784, seven years before Galvani's publication of *De viribus electricitatis*, electrical stimulation of muscle was common enough for Mauduyt to remark in one of his publications that it was a subject too well known to require documentation (Hoff, p. 163).

An electrical component to fertilization?

In fact, electricity was so much on the minds of biologists of the day that it was at this time that we find what may be the earliest reference to the possibility that it could be a driving force in development. In August 1778, Charles Bonnet, in a postscript to one of his letters to Lazzaro Spallanzani, speculated that there might be an electrical aspect to fertilization. He noted *la belle expérience de Monsieur Achard* (a German physicist principally interested in heat), in the course of which Achard had somehow been able to substitute electricity for the heat of an incubator to hatch chicken embryos, and Bonnet proposed that it should be possible to substitute "electrical fluid" for seminal fluid to fertilize eggs (Castellani, 1971, p. 370). This idea itself had a long incubation; it was not until the development of intense interest in artificial parthenogenesis in the early 1900s that Schücking (1903) and

Delage (1908a, b) added activation by an electric field to the many conditions that were reported to cause initiation of embryonic development without benefit of sperm. In Schücking's hands, one minute of direct current, delivered from two chromic acid voltaic cells, activated many (*zahlreicher*) starfish eggs to develop parthenogenetically, although he found that periods of two and five minutes of shock caused abnormal development and even death. Delage (1908a) imposed a static field across sea urchin eggs by connecting one lead from a 15-volt battery to the mixture of isotonic sodium chloride, isotonic sucrose, and seawater, in which the eggs were immersed, and the other lead to a tin plate cemented to the underside of the mica plate forming the base on which the eggs settled in their chamber. Development to pluteus larvae was seen after a treatment consisting of thirty minutes with the anode connected to the solution, followed by seventy-five minutes with the polarity reversed (Delage, 1908a). A cautious investigator, Delage (1908b) would not rule out entirely the possibility that these results were attributable to tiny (0.75 microampere) leakage currents' changing the pH of the solution in which the eggs were immersed, even though the amounts of acid and alkali produced by these currents were far less than those he found effective in experiments in which he caused parthenogenetic development by deliberately changing the pH of the solution in which the eggs were immersed.

The sequel to these experiments is that it ultimately was discovered, first in a tentative way by Péterfi and Rothschild in 1935 and with more certainty by Tyler and his colleagues in 1956, that there is indeed an electrical component to fertilization. In 1976, nearly two hundred years after Bonnet's speculation, Laurinda Jaffe reported her discovery that these changes provide a fast block to polyspermy, fertilization by more than one sperm. Immediately after fertilization of sea urchin eggs, the electric potential of the inside of the egg membrane rapidly changes from about -70 millivolts to as much as $+20$ millivolts inside positive. The probable connection between imposition of an electric field and parthenogenetic activation is that the Ca^{2+} channels that mediate a component of the rapid changes in electric potential are themselves dependent on the potential across the membrane (see, e.g., Jaffe, Gould-Somero, and Holland, 1979). Therefore, the fields imposed by Schücking and by Delage could well have changed the electric potential across the egg membrane sufficiently to cause a Ca^{2+} influx. This probably would be sufficient to initiate development, since there is a considerable body of evidence supporting the hypothesis that a Ca^{2+} influx is sufficient to activate the egg (reviewed by Jaffe, 1980).

Bioelectricity, nerve, and muscle

Bonnet's reference to *le fluide électrique* emphasizes a problem that was being wrestled with late in the eighteenth century (Hoff, 1936, p. 163). The prevailing notion of how nerves conducted their message to muscles was that nerves contained a highly movable fluid that could be set in motion by the will. At the middle of the century, Bertrand had his doubts: Nobody had ever seen this fluid, even with a microscope, and nerves are not hollow. With the growth of knowledge of electricity, it was becoming clear that this form of energy could well be the means by which nerve and muscle operate. Enter Luigi Galvani, who literally galvanized the world of science with his experiments on frog nerve and muscle. There were three of these experiments, the first two being published in his landmark article *De viribus electricitatis* in 1791 (Figure 10.2). With these two experiments, Galvani believed that he had demonstrated the existence of "animal electricity." Allesandro Volta, however, after closely examining Galvani's accounts of his experiments, realized that the muscles of Galvani's frogs were twitching from other sources of electricity: in the first experiment, from a static electricity generator and, in the second, from the voltage generated by contact between two joined, dissimilar metals and the preparation; Volta said so in print in 1792 (Fulton and Cushing, 1936, p. 243; Geddes and Hoff, 1971, pp. 39–44; Hoff, 1936, pp. 157–9). From this latter realization Volta proceeded to develop the battery and, in the course of his exchanges with Galvani, stimulated the third experiment, published anonymously in 1794, probably by the rather shy Galvani but perhaps with the help of his anything but shy nephew Giovanni Aldini. Here, muscle contraction was stimulated without any metals or static electricity generators, but by what Nobili later, in 1828, showed to be the injury potential of the cut spinal cord of the frog (Fulton and Cushing, pp. 245–7; Geddes and Hoff, pp. 44–5; Hoff, p. 160).

It is scarcely possible to overestimate the effect of Galvani's and Volta's experiments on the world of science and medicine. They ignited an intense interest in the role of electricity in physiology, medicine, and development that has been sustained to this day.

"Widespread irrational use of galvanism and static electricity"

The ill-advised enthusiasm of Galvani's nephew, Giovanni Aldini, was at least in part responsible for giving new impetus to the "widespread irrational use of galvanism and static electricity" (Stainbrook, 1948,

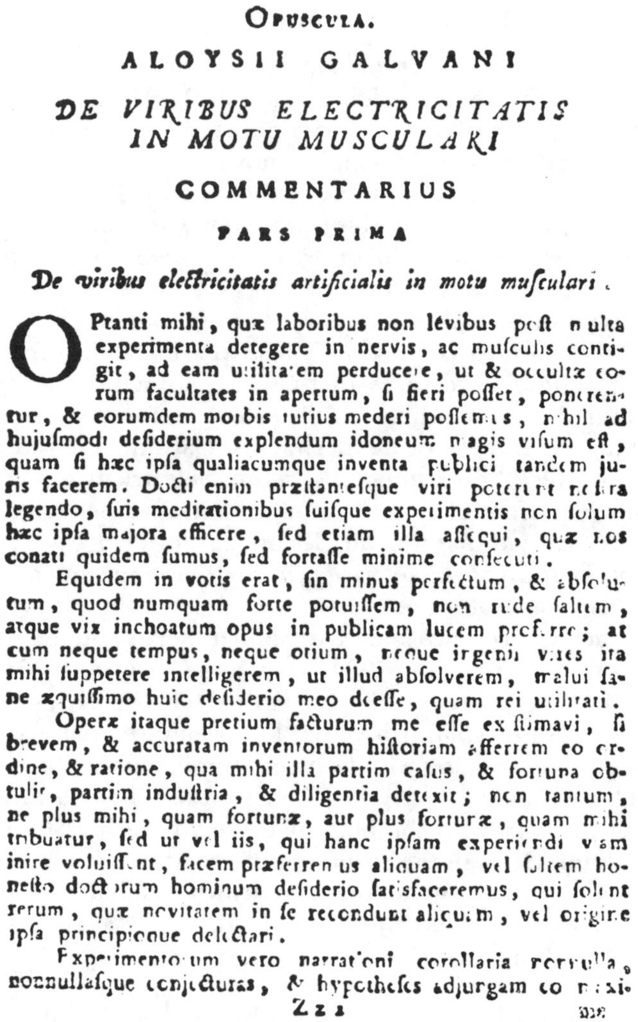

OPVSCVLA.

ALOYSII GALVANI

DE VIRIBUS ELECTRICITATIS
IN MOTU MUSCULARI

COMMENTARIUS

PARS PRIMA

De viribus electricitatis artificialis in motu musculari .

OPtanti mihi, quæ laboribus non lévibus peſt nulta
experimenta detegere in nervis, ac muſculis conti-
git, ad eam utilitatem perducere, ut & occultæ co-
rum facultates in apertum, fi fieri poſſet, poneren-
tur, & eorumdem morbis tutius mederi poſſemus, nihil ad
hujuſmodi deſiderium explendum idoneum nagis viſum eſt,
quam fi hæc ipſa qualiacumque inventa publici tandem ju-
ris facerem. Docti enim præſtantesque viri poterint reſra
legendo, ſuis meditationibus ſuiſque experimentis ncn ſolum
hæc ipſa majora efficere, ſed etiam illa aſſequi, quæ nos
conati quidem ſumus, ſed fortaſſe minime conſecuti .

Equidem in votis erat, fin minus perfectum, & abſolu-
tum, quod numquam forte potuiſſem, ncn rude ſaltem,
atque vix inchoatum opus in publicam lucem prof.rre; at
cum neque tempus, neque otium, reoue ingenii vires ita
mihi ſuppetere intelligerem, ut illud abſolverem, malui ſa-
ne æquiſſimo huic deſiderio meo deeſſe, quam rei utilitati .

Operæ itaque pretium facturum me eſſe ex ſtimavi, fi
brevem, & accuratam inventorum hiſtoriam afferrem eo or-
dine, & ratione, qua mihi illa partim caſus, & fortuna ob-
tulit, partim induſtria, & diligentia detexit; ncn tantum,
ne plus mihi, quam fortunæ, aut plus forturæ, quam mihi
tribuatur, ſed ut vel iis, qui hanc ipſam experiendi vam
inire voluiſſent, facem præferren us aliquam, vel ſaltem ho-
neſto doctorum hominum deſiderio ſatisfaceremus, qui ſolent
rerum, quæ novitatem in ſe recondunt aliquam, vel origire
ipſa principioque delectari .

Experimentorum vero narrationi corollaria nerrulla,
nonnullaſque conjecturas, & hypetheſes adjurgam co r:si.

Z z z   me

Figure 10.2. A reproduction of the first page of the original 1791 publication of
Galvani's seminal paper that led to an understanding of "animal electricity," which
appeared in the *Proceedings* of the University of Bologna. (Plate 18 in Fulton and
Cushing, 1936. Reproduced with the publisher's permission.)

p. 157) in medicine that began late in the eighteenth century and
continued with deplorable enthusiasm until at least the beginning of
the twentieth (Fulton and Cushing, p. 249; Stainbrook, pp. 157ff.).
In his two-volume essay *Sur le galvanisme,* published in 1804, Aldini

Figure 10.3. An electric air-bath treatment, delivered from the static electricity generator in the background. One electrode was grounded; the other, connected to the patient's clothing. Sparks could be drawn (in this instance, from the elbow) by holding a grounded electrode (see Figure 10.5) close to the desired part of the patient's body (Geddes, p. 5–6). A positive air bath supposedly produced a feeling of exhilaration. (Figure 2–3 in Geddes, 1984. Reproduced with the publisher's permission.)

seized the limelight and spoke of spectacular demonstrations of the power of electricity. He described his trip to London to apply the voltage from a battery to the muscles of an executed prisoner, eliciting violent muscular contractions "almost to give the appearance of reanimation" (Geddes, 1984, pp. S-9, S-10), and efforts to cure insanity at an infamous Paris insane asylum, the Salpetrière. He advocated galvanism as a treatment for "melancholy madness," although, with uncharacteristic caution, he warned that it was dangerous for treating "raving madness" (without indicating to whom). This approach to treating mental disease has continued to the present day in one form or another, often arousing intense controversy, but in specific instances to good effect (Stainbrook, pp. 169ff.).

One notorious example of the medical charlatanism of the time is the entrepreneurship of Dr. Elisha Perkins, with his "metallic points to relieve pain" (Haggard, 1936, pp. 137ff.). These V-shaped devices,

Figure 10.4. A negative breeze treatment. The negative electrode completing the circuit, placed just above the recipient's head, is not shown. The negative breeze was considered helpful in dealing with insomnia, migraine, baldness, and the early stages of kidney disease (Reynolds, p. 6). The positive breeze, with the electrodes reversed, was used for the later stages of kidney disease, if the early treatment was not effective. (Figure 1–1 in Reynolds, 1971. Courtesy of C. C. Thomas, Publisher, Springfield, Illinois.)

consisting of two joined, tapered rods, one of iron, the other of brass, were peddled at a hundredfold markup in the United States and England from 1795 on. Static electricity generators abounded; the "electric air bath" was popular (Figure 10.3; Geddes and Hoff, p. S-6), as was the "negative breeze" (Figure 10.4; Reynolds, 1971, p. 7). Batteries were connected to people through a variety of electrodes (Figure 10.5) for "the abatement of uneasy sensations; composure or

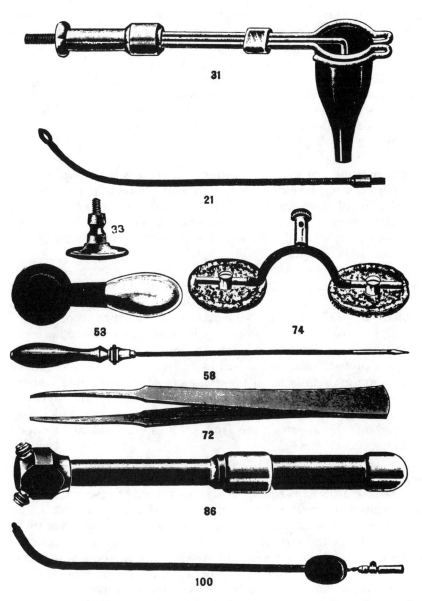

Figure 10.5. Various devices for making electrical contact during electrotherapeutic treatment, and other electrotherapeutic accessories. (Figure 1–3 in Reynolds, 1971. Courtesy of C. C. Thomas, Springfield, Illinois.)

exhilaration of the animal spirits; better appetite and improved digestion; sound and refreshing sleep; and increased discharges from the
bladder and bowels" (Geddes, 1984, p. S-11).

Having a battery in the operating room proved to be genuinely
valuable, however. In the late nineteenth century, chloroform was
widely used for anesthesia, often with the undesired side effects of
respiratory depression and cardiac arrest. Briefly connecting the battery to such unfortunate patients often had the welcome effect of
stimulating the diaphragm and the heart to reinitiate their function
(Geddes, 1984, p. S-13).

The invention of the "inductorium" by DuBois-Reymond in 1848
introduced a third mode of electrotherapy, "faradic" stimulation, a
rapid train of brief pulses of current, to the modes of static and
galvanic electricity. This mode was considerably more dangerous than
the first two; it tended to *cause* cardiac fibrillation rather than cure it
(Geddes, 1984, p. S-21).

A vigorous area of electrotherapeutics grew up in the latter part of
the nineteenth century, until the excesses of its practitioners caused it
to be discredited early in the twentieth (Figure 10.6; Longo, 1986,
p. 385). Largely in reaction to the excesses of the surgeons of the day,
faradic stimulation was advocated as a humane alternative to the treatment of uterine fibroids and pelvic inflammatory disease. As such,
that was probably not such a bad idea. From this, however, faradic
stimulation grew to be used for treatment of an ill-advisedly broad
range of gynecological problems, including dysmenorrhea, infertility,
hysteroneurosis, and menopause. When finally it became widely recognized that this approach was fruitless, it rather suddenly dropped
from the scene.

The point of dwelling at such length on the history of electrical
approaches to medicine is that it has been against this checkered background that a perhaps equally checkered record of research on electrical aspects of development and regeneration has accrued. As an
investigator in this area, I have had the distinct impression that there
has been, and I think properly continues to be, a dogged skepticism
about claims for the relevance of steady ionic currents and steady
electric fields to development in general, and to regeneration in particular. I think that, at least in part, this attitude has its roots not only
in the sometimes questionable developmental studies carried out in
the past by erstwhile developmental bioelectricians but also in the
questionable medical applications of electrical procedures that have
flourished in the past.

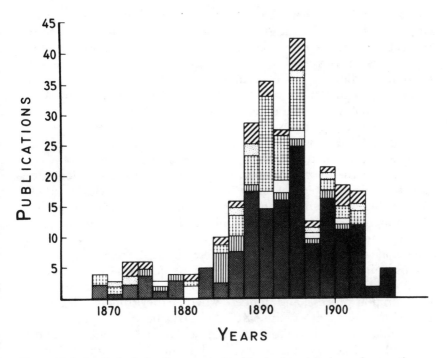

Figure 10.6. Graph depicting frequency of articles in American journals reporting electrotherapeutic treatment of various obstetric and gynecological diseases between 1868 and 1906. By 1910, the electrotherapeutic approach to these diseases had virtually been abandoned. (Figure 1 of Longo, 1986. Reproduced with the publisher's permission.)

Bioelectricity and development

What, then, about the bioelectrical research on development and regeneration that was going on in the days of such exuberant interest in electricity? We have already considered Bonnet's speculation about electricity and egg activation. It is possible to identify at least four other paths of bioelectrical research on animal development, some of them converging, that can be traced from the middle of the nineteenth century to modern times.

Wound currents

The first of these is perhaps the most respectable: Sometime before 1849, the great physiologist Emil DuBois-Reymond, discovered wound currents in one of his many investigations into animal electricity. He found that when he made a cut in one of his fingers, he could cause the deflection of a galvanometer by putting this finger and a contralateral unwounded finger into the circuit (DuBois-Reymond, 1857). His measurements indicated that about 1 microampere of current was flowing out of such skin wounds. In 1849, this experiment was reported to the Royal Astronomical Society by W. G. Lettsom, Esq., with the assurance that he was able to repeat it. (Such repeatability, as we shall soon see, was not to be taken for granted in those days – for that matter, in these days, as well.) Sixty years later, Amedeo Herlitzka, a student of DuBois-Reymond, reported the continuation of these experiments on wound currents. He called attention to their possible relevance to regeneration (Herlitzka, 1910): "Durch diesen Gedankengang geleitet, habe ich mir die Frage gestellt, ob es möglich wäre nachzuweisen, ob bei dem Anfang eines Regenerationsprozesses elektrische Vorgänge eine Rolle spielen, und zwar ob diese nach dem Abtragen eines Gewebeteils die Zellteilung einleiten." ("This train of thought led me to pose the question of whether it is possible to show that electric currents play a role at the onset of the regeneration process, and, indeed, whether they initiate cell division following removal of a tissue part.") Herlitzka here was using the term *Regenerationsprozess* in quite a broad sense. He perhaps more properly should have said "wound healing" rather than "regeneration," but it seems significant that he used the term. As we shall soon see, by this time, in the United States, Albert Mathews had reported his experiments that initiated a long series of investigations on the relations between bioelectricity and epimorphic regeneration. Herlitzka was aware of this work and discussed it in relation to his own.

Herlitzka observed that as his skin wounds healed, the currents decreased. We know now that this is because the healing seals the electrical leak produced by the wound. Burr made essentially the same observation nearly thirty years later (Figure 10.7) by measuring the change in electric potential between normal skin and a wound made in the skin as wound healing progressed (Burr, Harvey, and Taffel, 1938). Nearly fifty years after Burr, with more refined instrumentation, we also were able to see that as a wound heals, wound currents diminish (McGinnis and Vanable, 1986b). A repeatable observation, indeed.

The force driving these currents, from the epidermis of the skin, was being studied rather intensively late in the nineteenth and early in the twentieth century. In 1885, Hermann and von Gendre reported that fish skin maintains a voltage across itself, inside positive, and in 1894, Reid reported the same result with chicken skin. In 1901,

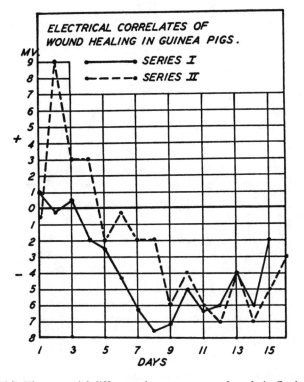

Figure 10.7. The potential difference between a wound made in flank skin of the guinea pig and the surface of unwounded skin as healing proceeds. As the wound closes, the electrode at the wound site becomes electrically isolated from the space beneath the skin, which is positive with respect to the skin surface, and so the potential difference between the two electrodes diminishes. It is interesting that during later phases of wound healing the wound electrode becomes negative with respect to the electrode over unwounded skin. This may be due to the lack of hair on the newly healed skin, possibly enabling this region to develop a greater transepithelial potential than adjacent hairy areas. (See Barker et al., 1982.) (Figure 1 in Burr et al., 1938. Reproduced with the publisher's permission.)

Augustus D. Waller, son of the Augustus V. Waller of Wallerian degeneration fame, reported the first of a series of studies on skin of amphibians and mammals (Waller, 1901a, b; 1902a, b) that confirmed the fact that skin acts as a battery. He also found that when skin is stimulated electrically, it generates a train of spikes that he imaginatively termed "blaze currents." Figure 10.8 shows a few of these in the frog. These intriguing spikes have been seen since, by Finkelstein in the 1960s (Finkelstein, 1964), and more recently by us (McGinnis and Vanable, 1986b), but their significance remains elusive. The relevance

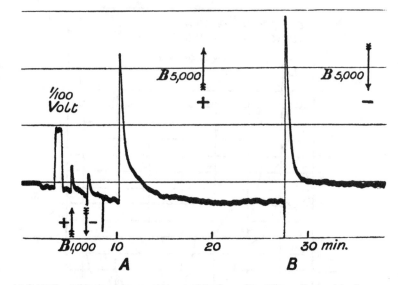

Figure 10.8. Waller's blaze currents, measured in frog skin. The spike at *A* is the response to stimulation when the cathode was placed on the surface of the skin; the spike at *B* is the response when the anode was placed on the skin surface. Similar observations were made with cat skin (Waller, 1901b) and human skin (Waller, 1902b). (From Waller, 1901a. Reproduced with the publisher's permission.)

of steady skin potentials to wound healing and to limb regeneration has, however, remained an active and, I think, proper object of investigation to this day (McGinnis and Vanable, 1986b; Hearson, Eltinge, and Vanable, 1988; Borgens, 1989b; Vanable, 1989).

Effects of electric fields on cells

Just forty years after DuBois-Reymond's seminal and robust studies of the skin's ability to develop an electric potential across itself, the early investigation of the effect of electric fields on cells was beginning. In 1892, Dineur reported to the Belgian Microscopical Society that in the living frog, leucocytes migrate toward the anode of an imposed electric field, unless they had gathered in response to an inflammation, in which case he found that they migrate toward the cathode. The field strengths he used were quite physiological, comparable to those generated by normal tissues; he was far ahead of his time. Eleven years later, at Harvard, R. S. Lillie (1903) carried out similar studies, this time in vitro, while directly observing the cells through his microscope. Not to be outdone, Sven Ingvar, working at Yale in 1920 with

R. G. Harrison's cultured neural tube preparation, reported (without giving many details, unfortunately) that outgrowing nerve fibers oriented themselves parallel to an imposed electric field. Studies of the effect of imposed fields on the migration and growth of cultured cells have, of course, continued to the present (Robinson, 1985). It is a fruitful and exciting area, and one vital to the understanding of how electricity might affect development and regeneration.

Electrical correlates of polarity

Another significant area of inquiry into the role of bioelectricity and development has addressed the question of the polarity of eggs and ooplasmic segregation. Around the turn of the century, Ida Hyde, landlocked most of the year at the University of Kansas, at times during the summer visited marine laboratories, both the Stazione Zoologica di Napoli, in Italy, and the Marine Biological Laboratory at Woods Hole, Massachusetts. As well as showing good judgment with regard to the use of her summers, Ida must have been a patient and meticulous person: Using an inherently finicky capillary electrometer, she was able to measure a potential difference between the animal and vegetal poles of turtle and fish eggs that, of course, reflected their animal-vegetal polarity (Hyde, 1905). Nearly forty years later, at Yale University Medical School, H. S. Burr (1941), apparently keeping in touch with R. G. Harrison on the other side of town (Northrop and Burr, 1937), reported some remarkable measurements on frog embryos that generally confirmed and extended Hyde's observations. Burr had a much better instrument, a high-impedance voltmeter he had developed with his colleagues Lane and Nims (1936). After using it, they reported that the animal pole of the frog egg is negative with respect to the vegetal pole and that the neuraxis of the developing embryo is predicted by the potential difference between the animal pole and a point on the egg's equator that becomes the anterior edge of the neural plate. They reported that this potential difference is somewhat greater than the potential differences between the animal pole and other points on the equator (Figure 10.9). These observations bear repeating, however. Robinson (1979), using the ultrasensitive vibrating electrode, was able to measure currents of the order of 1 microampere per square centimeter entering the animal pole of immature *Xenopus* oocytes and leaving the vegetal pole, which indicates that in the immature oocyte the animal pole is negative with respect to the vegetal pole, as Burr and his colleagues reported. Upon maturation, currents could not be detected at any point on the surface

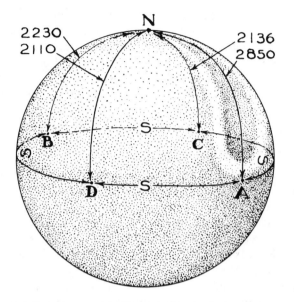

Figure 10.9. Burr's measurements of electrical polarity of the frog's "egg" (probably a late gastrula), indicating its animal–vegetal polarity and predicting its future neuraxis. Values are in microvolts. N is the animal pole; S indicates statistically significant differences in potential between equatorial points. (Figure 2 in Burr, 1941. Reproduced with the publisher's permission.)

of the egg (Robinson, p. 838), but during fertilization (Kline and Nuccitelli, 1985) and cleavage (Kline, Robinson, and Nuccitelli, 1983) transient currents have been measured. It is not likely that Burr could have measured the potentials associated with these transient currents of fertilization and early cleavage. This would raise a serious question about the accuracy of his main body of data, were it not for the fact that careful reading of his report reveals that most of his experiments with "eggs" were actually done with embryos at stages "prior to the development of the primary axis of the embryo as seen in the appearance of the medullary plate" (p. 278), most likely late gastrulae. Current measurements have not been reported for this stage of amphibian embryo, but in neurulae strong outcurrents can be measured leaving the blastopore (Robinson and Stump, 1984). It therefore is quite likely that there would also be outcurrents from the blastopore of a late gastrula. This direction of current would be consistent with the polarity of the potential between animal and vegetal poles that Burr and his colleagues reported. They did report, however, without giving much detail, that "typical field patterns" were also observed in unfertilized eggs and in fertilized but uncleaved zygotes. These claims, then, need to be reexamined.

Bioelectricity and epimorphic regeneration

The most distinct beginning I have been able to find to the study of bioelectricity and epimorphic regeneration lay in the ideas developed at the University of Chicago in C. M. Child's day. In the early 1900s, Child and his colleagues were immersed in the investigation of the role of metabolic gradients in developing and regenerating animals (see Chapter 12 in the present volume). Albert Mathews (1903) pioneered in relating the polarity of electric potential differences to these gradients and to regeneration, with electrical measurements on segments of the colonial coelenterate *Parypha*. He reported that the polyp surface of these segments, where regeneration is faster than at the stolon surface, invariably was about 5 millivolts electronegative with respect to the stolon surface. He even made preliminary attempts to determine whether this potential difference is relevant to the polarity of regeneration by imposing countercurrents on the segments, and he reported that in "many cases" he was able to change which end regenerated a new polyp.

None other than T. H. Morgan (Morgan and Dimon, 1904; see also Chapter 9 in the present volume) became interested in this line of investigation when making electrical measurements of earthworm segments. This study may well have been one of the factors that drove Morgan from regeneration to genetics; he tried assiduously but was unable to find any correlation at all between the potentials he measured in these segments and their regenerative capacity.

Still, many measurements of potentials, reliable and unreliable, were being made. In Italy, Vialle (1916), stimulated by Herlitzka's studies of wound healing, investigated the role of electric polarity and nerve regeneration. He cut guinea pig sciatic nerves and measured the potential difference between the proximal nerve trunk and its cut end. Interestingly, after much convoluted discussion, he convinced himself that the cut end was positive with respect to the stump. These results matched the polarity found by Herlitzka (and others) in skin cuts but were just the opposite of what the true polarity must have been. In similar studies of potential differences related to tadpole injuries (1921), he was back on track, with a polarity that matched both reality and Herlitzka's results.

Meanwhile, Hyman and Bellamy (1922), following in the footsteps of Mathews at the University of Chicago, doggedly collected a series of

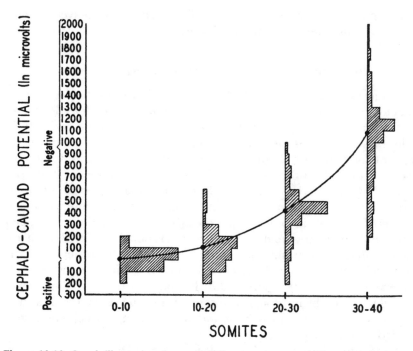

Figure 10.10. Graph illustrating the small differences in potential between the anterior and posterior ends of developing chicken embryos, one of several contributions by Burr and his colleagues providing electrical correlates to C. M. Child's ideas about polarity and development. The histograms for each point indicate the frequency distribution of values of individual measurements. (Figure 1 in Burr and Hovland, 1937a. Reproduced with publisher's permission.)

potential measurements in a wide range of species, from coelenterates to tadpoles, trying to correlate their observations with Child's metabolic gradient studies. Burr, with his powerful high-impedance voltmeter, continued this tradition by measuring potentials in developing chick and salamander embryos (Burr and Hovland, 1937a, b); his results with the chick are shown in Figure 10.10. Burr also measured the potential differences along segments of *Obelia* (Burr and Hammett, 1939) and reported without comment a polarity opposite to that determined by Mathews (1903). He certainly was aware of the work of Mathews; he had cited it in his earlier publications (Burr and Hovland, 1937a, b). He missed a good opportunity; as it turns out, it was Mathews who was mistaken.

At about this time, Morgan's despair notwithstanding, the indefatigable Elmer Lund began his long and largely fruitful series of investigations of bioelectricity and development (summarized in Lund,

1947). Lund began (1921) with studies on the regeneration of segments in the marine hydroid *Obelia,* attacking right away the question of relevance posed earlier by Mathews. With a well-designed apparatus, he imposed electric fields on segments of *Obelia* to see if he could influence the polarity of their regeneration (Figure 10.11). Segments of *Obelia* typically regenerate a hydranth from each end – the distal (hydranth) end develops first, but eventually the proximal end develops as well (Figure 10.12). But when Lund imposed electric fields across the segments, they regenerated a hydranth at only one end, the one closer to the anode. At the cathode end, hydranth regeneration was inhibited (Figure 10.12).

This is a striking result, at first glance, but there was a problem: Normally the hydranth develops earliest at what Lund at first (1922), perhaps misled by Mathews, had determined to be the cathode end. Paradoxically, in these experiments (Lund, 1921) hydranth formation was *inhibited* at the cathode end. This difficulty, which Lund did not discuss in his first articles (1921, 1922), was remedied by his later experiments with improved equipment (1925), for he realized that his earlier measurements, made with inadequate equipment, had resulted in an incorrect assignment of polarity for *Obelia* segments. With the correct assignment of polarity, Lund's experiments became a beautiful demonstration of the control of regeneration by imposed electric fields.

Lester Barth (1934a, b) continued the work on colonial hydroid regeneration that was begun by Lund, with four different species and with considerable success. In each case, the polarity of the effect of the imposed fields was consistent with the inherent polarity he was able to measure in these species, and he, like Lund, was able to reverse the polarity of regeneration by imposing fields that countered the inherent polarity.

Some years later, from the laboratory of Alberto Monroy, came the work that ultimately gave rise to studies on bioelectricity and amphibian limb regeneration that have been conducted in recent years. In 1941, Monroy reported that an electric potential exists between the surface of an amputated salamander stump and the skin of the stump. In retrospect, this finding is not surprising, in view of the work of DuBois-Reymond and Herlitzka discussed earlier; nonetheless, this observation can be recognized as the catalyst for present-day studies of bioelectricity and amphibian limb regeneration. Robert Becker (1960), the Roses (1974), and Bernard Lassalle (1974a, b; 1979, 1980) were inspired by this study, and their work in turn inspired the remarkable experiments of S. D. Smith (1974), in which current

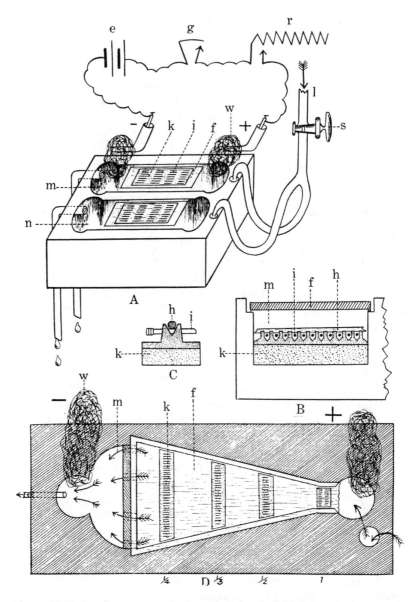

Figure 10.11. Lund's apparatus for imposing electric fields across regenerating segments of *Obelia* stems. In *A*, current is delivered via bridges (to reduce or eliminate the possibility of electrode products' confounding the results) to the far chamber; the adjacent chamber contains control segments, which receive no current. In *B* and *C*, details of the grooved cork in which the segments were held are shown. *D* is a drawing of Lund's ingenious apparatus for varying the current density, and hence the drop in electric potential, across the regenerating segments. (Figure 1 in Lund, 1921. Reproduced with the publisher's permission.)

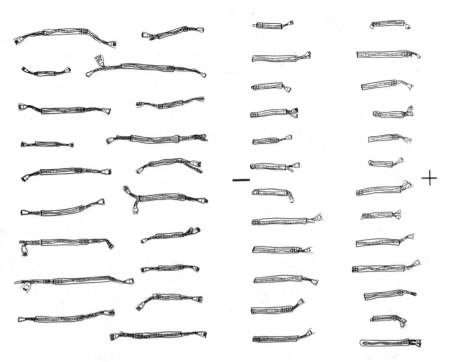

Figure 10.12. The effect of imposing electric fields across regenerating internodes of *Obelia*. On the left are the controls, showing the typical result of hydranth regeneration eventually occurring at both ends. On the right is shown regeneration in internodes across which an electric potential had been imposed. Hydranth regeneration at the end facing the cathode has been inhibited, regardless of whether this end is basal (rings in the perisarc) or apical (no rings). This result is consistent with the fact that during normal regeneration hydranth formation occurs earliest at the anode end of the internode (Lund, 1925. Plates IA and IB of Lund, 1921. Reproduced with the publisher's permission.)

drawn through the amputation stumps of adult frogs caused a great deal more to develop from those stumps than ever does without such treatment. Lionel Jaffe attended the meeting of the New York Academy of Sciences at which Smith presented this work and, returning full of enthusiasm for it, persuaded Richard Borgens, who had just arrived as a new graduate student, and me to collaborate with him in extending it. The research begun then at Purdue University has continued to this day with other collaborators, particularly Lester Hearson, Charles Thornton's first graduate student (Thornton, 1968), who has provided an excellent pair of hands, stimulating ideas, and a clear perspective on the process of regeneration (Borgens, Vanable, and Jaffe, 1979; McGinnis and Vanable, 1986a, b; Hearson, Eltinge, and Vanable, 1988).

What does all this tell us about bioelectricity and development and regeneration? One important message is that, like bioelectricity and medicine, some ideas and experiments have held up; many have not. Universal gynecological electrotherapy did not survive, but such procedures and devices as cardiac pacemakers, electrocautery, cardiac defibrillation, and even electroshock therapy persist. The early measurements of potential in developing and regenerating systems were done only with great difficulty, and the results were uncertain: Some were correct, at least qualitatively; others were not. These were measurements of surface potentials, and most of these almost certainly are not relevant to development and regeneration in multicellular systems. Today, internal potentials, the relevant ones, are being measured (Barker, Jaffe, and Vanable, 1982; McGinnis and Vanable, 1986b; Chiang and Vanable, 1989), and it turns out that these potentials are large enough to influence cell behavior. Tests of the relevance of these fields for regeneration and development continue to be made in many laboratories around the world (For reviews, see Borgens, 1989b, c; Borgens and McCaig, 1989; Jaffe, 1989; McGinnis, 1989; and Robinson, 1989), but still not to everyone's satisfaction (including, in many cases, my own).

This area of inquiry as been a stormy and discouraging one, full of difficulties and controversy. But it still holds promise, I would submit, for providing useful information about development and regeneration. It is hoped that history will record that the present era of vigorous experimentation will have provided a firm foundation upon which the ultimately definitive work will rest.

ACKNOWLEDGMENTS

I am indebted to Charles E. Dinsmore for calling my attention to Charles Bonnet's speculation concerning an electrical component of fertilization, and to K. R. Robinson for helpful suggestions that improved several sections of this essay.

REFERENCES

Barker, A. T., L. F. Jaffe, and J. W. Vanable, Jr. 1982. The glabrous epidermis of cavies contains a powerful battery. *Am. J. Physiol.*, 242: R358–R366.
Barth, L. G. 1934a. The effect of constant electric current on the regeneration of certain hydroids. *Physiol. Zool.*, 7: 340–64.
    1934b. The direction and magnitude of potential differences in certain hydroids. *Physiol. Zool.*, 7: 365–99.

Becker, R. O. 1960. The bioelectric field pattern in the salamander and its simulation by an electronic analog. *IRE Trans. Med. Electronics ME*, 7: 202–7.

Borgens, R. B. 1989a. Introduction. In R. B. Borgens, K. R. Robinson, J. W. Vanable, Jr., and M. E. McGinnis, *Electric Fields in Vertebrate Repair: Natural and Applied Voltages in Vertebrate Regeneration and Wound Healing*. Liss, New York, pp. ix–xxi.

1989b. Natural and applied currents in limb regeneration and development. In R. B. Borgens, K. R. Robinson, J. W. Vanable, Jr., and M. E. McGinnis, *Electric Fields in Vertebrate Repair: Natural and Applied Voltages in Vertebrate Regeneration and Wound Healing*. Liss, New York, pp. 27–75.

1989c. Artificially controlling axonal regeneration and development by applied electric fields. In R. B. Borgens, K. R. Robinson, J. W. Vanable, Jr., and M. E. McGinnis, *Electric Fields in Vertebrate Repair: Natural and Applied Voltages in Vertebrate Regeneration and Wound Healing*. Liss, New York, pp. 117–170.

Borgens, R. B., and C. D. McCaig. 1989. Endogenous currents in nerve repair, regeneration, and development. In R. B. Borgens, K. R. Robinson, J. W. Vanable, Jr., and M. E. McGinnis, *Electric Fields in Vertebrate Repair: Natural and Applied Voltages in Vertebrate Regeneration and Wound Healing*. Liss, New York, pp. 77–116.

Borgens, R. B., J. W. Vanable, Jr., and L. F. Jaffe. 1979. Bioelectricity and regeneration. *BioScience*, 29: 468–74.

Burr, H. S. 1941. Field properties of the developing frog's egg. *Proc. Natl. Acad. Sci. USA*, 27: 276–81.

Burr, H. S., and F. S. Hammet. 1939. A preliminary study of electric correlates of growth in *Obelia geniculata*. *Growth*, 3: 211–20.

Burr, H. S.; S. C. Harvey; and M. Taffel. 1938. Bio-electric correlates of wound healing. *Yale J. Biol. Med.* 11: 103–7.

Burr, H. S., and C. I. Hovland, 1937a. Bio-electric potential gradients in the chick. *Yale J. Biol. Med.*, 9: 247–58.

1937b. Bio-electric correlates of development in *Amblystoma*. *Yale J. Biol. Med.*, 9: 541–49.

Burr, H. S.; C. T. Lane; and L. F. Nims. 1936. A vacuum tube microvoltmeter for the measurement of bioelectric phenomena. *Yale J. Biol. Med.*, 9: 65-76.

Castellani, C. 1971. *Lettres à le monsieur l'abbé Spallanzani de Charles Bonnet*. Critical edition. Episteme Editrice, Milan.

Chiang, M., E. J. Cragoe, Jr., and J. W. Vanable, Jr. 1989. Electrical fields in the vicinity of small wounds in *Notophthalmus viridescens* skin. *Biol. Bull.*, 176(S): 179–83.

Delage, Y. 1908a. La parthénogenese expérimentale par les charges électriques. *C. R. Acad. Sci.*, 147: 553–7.

1908b. Sur le mode d'action de l'électricité dans la parthénogenèse électrique. *C. R. Acad. Sci.*, 147: 1372–8.

Dineur, E. 1892 Note sur la sensibilité des leucocytes à l'électricité. *Bull. Séances Soc. Belge Microsc.* (Brussels), 18: 113–18.

DuBois-Reymond, E., 1848. *Untersuchung über thierische Elektrizität*. G. Reimer, Berlin. (Cited from Hoff, 1936.)

Finkelstein, A. 1964. Electrical excitability of frog skin and toad bladder. *J. Gen. Physiol.*, 47: 543–65.

Fulton, J. F., and H. Cushing. 1936. A bibliographic study of the Galvani and the Aldini writings on animal electricity. *Ann. Sci.*, 1: 239–68.

Geddes, L. A. 1984. A short history of the electrical stimulation of excitable tissue. *Physiologist* (suppl.), 27: S1–S47.

Geddes, L. A., and H. E. Hoff. 1971. The discovery of bioelectricity and current electricity: The Galvani–Volta controversy. *IEEE Spectrum*, 8: 38–46.

Haggard, H. W. 1936. The first published attack on Perkinism: An anonymous eighteenth century poetical satire. *Yale J. Biol. Med.*, 9: 137–53.

Hearson, L. L., E. M. Eltinge, and J. W. Vanable, Jr. 1988. Do stump fields promote the nerve growth that is necessary for limb regeneration? *Proceedings. Sixth M. Singer Symposium*, ed. S. Inoue, T. Shirai, M. W. Egar, S. Aiyama, et al. Okada Printing and Publishing, Maebashi, Japan, pp. 211–20.

Herlitzka, A. 1910. Ein Beitrag zur Physiologie der Generation. Elektrophysiologische Untersuchungen. *Wilhelm Roux Archiv f. Entwicklungsmech.*, 30: 126–58.

Hermann, L., and B. von Gendre. 1885. Über eine electromotorische Eigenschaft des bebrüteten Hühnereies. *(Pflüger's) Arch. f. ges. Physiol.*, 35: 34–5.

Hoff, H. E. 1936. Galvani and the pre-Galvanian electrophysiologists. *Ann. Sci.*, 1: 157–72.

Hyde, I. H. 1905. Differences in electrical potential in developing eggs. *Am. J. Physiol.*, 12: 241–75.

Hyman, L. H., and A. W. Bellamy. 1922. Studies on the correlation between metabolic gradients, electrical gradients and galvanotaxis. *Biol. Bull.*, 43: 313–47.

Ingvar, S. 1920. Reaction of cells to the galvanic current in tissue cultures. *Proc. Soc. Exp. Biol. Med.*, 17: 198–9.

Jaffe, L. A. 1976. Fast block to polyspermy in sea urchin eggs is electrically mediated. *Nature*, 261: 68–71.

Jaffe, L. A.; M. Gould-Somero; and L. Holland. 1979. Ionic mechanism of the fertilization potential of the marine worm, *Urechis caupo* (Echiura). *J. Gen. Physiol.*, 73:469–92.

Jaffe, L. F. 1980. Calcium explosions as triggers of development. *Ann. N.Y. Acad. Sci.*, 339: 86–101.

1989. Preface. *Biol. Bull.*, 176(S): 3–4.

Kellaway, P. E. C. 1946. The part played by electric fish in the early history of bioelectricity and electrotherapy. *Bull. Hist. Med.*, 20: 112–37.

Kline, D., and R. Nuccitelli. 1985. The wave of activation current in the *Xenopus* egg. *Dev. Biol.*, 111: 471–87.

Kline, D.; K. R. Robinson; and R. Nuccitelli. 1983. Ion currents and membrane domains in the cleaving *Xenopus* egg. *J. Cell Biol.*, 97: 1753–61.

Lassalle, B. 1974a. Charactéristiques des potentiels de surface des membres du triton *Pleurodeles watlii*, Michah. *C. R. Acad. Sci. (Paris), Series D*, 278: 483–6.

1974b. Origine épidermique des potentiels de surface du membre de triton, *Pleurodeles watlii*, Michah. *C. R. Acad. Sci. (Paris), Series D*, 278: 1055–8.

1979. Surface potentials and the control of amphibian regeneration. *J. Embryol. Exp. Morphol.*, 53: 213–23.

1980. Are surface potentials necessary for amphibian regeneration? *Dev. Biol.*, 75: 460–6.

Lettsom, W. G. 1849. Deflection of the magnetic needle by the act of volition. *Phil. Mag.*, 34: 543–5.

Licht, S. 1944. The history of electrodiagnosis. *Bull. Hist. Med.*, 16: 450–67.

Lillie, R. S. 1903. On differences in the direction of the electrical convection of certain free cells and nuclei. *Am. J. Physiol.*, 8: 273–83.

Longo, L. D. 1986. Electrotherapy in gynecology: The American experience. *Bull. Hist. Med,.* 60: 343–66.

Lund, E. J. 1921. Experimental control of organic polarity by the electric current. I. Effects of the electric current on regenerating internodes of *Obelia commissuralis. J. Exp. Zool.*, 34: 471–93.

1922. Experimental control of organic polarity by the electric current. II. The normal electrical polarity of *Obelia:* A proof of its existence. *J. Exp. Zool.*, 36: 477–94.

1925. Experimental control of organic polarity by the electric current. V. The nature of the control of organic polarity by the electric current. *J. Exp. Zool.*, 41: 155–90.

1947. *Bioelectric Fields and Growth*. University of Texas Press, Austin.

McGinnis, M. E. 1989. The nature and effects of electricity in bone. In R. B. Borgens; K. R. Robinson; J. W. Vanable, Jr., and M. E. McGinnis, *Electric Fields in Vertebrate Repair: Natural and Applied Voltages in Vertebrate Regeneration and Wound Healing*. Liss, New York, pp. 225–84.

McGinnis, M. E., and J. W. Vanable, Jr. 1986a. Wound epithelium resistance controls stump currents. *Dev. Biol.*, 116: 174–83.

1986b. Electrical fields in *Notophthalmus viridescens* limb stumps. *Dev. Biol.*, 116: 184–93.

Mathews, A. P. 1903. Electrical polarity in the hydroids. *Am. J. Physiol.*, 8: 294–9.

Monroy, A. 1941. Ricerche sulle correnti ellettriche derivabili dalla superficie del corpo di Tritone adulti normali e durante la rigenerazione degli arti e della coda. *Pubbl. Stat. Zool. Napoli*, 18: 265–81.

Morgan, T. H., and A. C. Dimon. 1904. An examination of the problems of physiological polarity and of electrical polarity in the earthworm. *J. Exp. Zool.*, 1: 331–47.

Northrop, F. S. C., and H. S. Burr. 1937. Experimental findings concerning the electrodynamic theory of life and an analysis of their physical meaning. *Growth*, 1: 78–88.

Péterfi, T., and V. Rothschild. 1935. Bio-electric transients during fertilisation. *Nature*, 135: 874–5.

Reid, E. W. 1894. Electromotive phenomena in non-secretory epithelia. *J. Physiol.*, 16: 360–7.

Reynolds, D. V. 1971. A brief history of electrotherapeutics. In D. E. Reynolds and A. E. Sjoberg, eds., *Neuroelectric Research: Electroneuroprosthesis, Electroanesthesia, and Nonconvulsive Electrotherapy*. Thomas, Springfield, Ill., pp. 5–12.

Robinson, K. R. 1979. Electrical currents through full-grown and maturing *Xenopus* oocytes. *Proc. Natl. Acad. Sci. USA*, 76: 837–41.

1985. The responses of cells to electrical fields: A review. *J. Cell Biol.*, 101: 2023–7.

1989. Endogenous and applied electrical currents: Their measurement and application. In R. B. Borgens, K. R. Robinson, J. W. Vanable, Jr., and M. E. McGinnis, *Electric Fields in Vertebrate Repair: Natural and Applied Voltages in Vertebrate Regeneration and Wound Healing.* Liss, New York, pp. 1-25.

Robinson, K. R., and R. F. Stump. 1984. Self-generated electrical currents through *Xenopus* neurulae. *J. Physiol.*, 352: 339–52.

Rose, S. M., and F. C. Rose. 1974. Electrical studies on normally regenerating, on x-rayed, and on denervated limb stumps of *Triturus. Growth*, 38: 363–80.

Schücking, A. 1903. Zur Physiologie der Befruchtung, Parthenogenese un d Entwicklung. *(Pflüger's) Archiv f. Ges. Physiol.*, 97: 58–97.

Smith, S. D. 1974. Effects of electrode placement on stimulation of adult frog limb regeneration. *Ann. N.Y. Acad. Sci.*, 238: 500–7.

Stainbrook, E. 1948. The use of electricity in psychiatric treatment during the nineteenth century. *Bull. Hist. Med.* 22: 156–77.

Thornton, C. S. 1968. Amphibian limb regeneration. *Adv. Morphogen.*, 7: 205–49.

Tyler, A., A. Monroy, C. Y. Kao, and H. Grundfest. 1956. Membrane potential and resistance of the starfish egg before and after fertilization. *Biol. Bull.* 111: 153–77.

Vanable, J. W., Jr. 1989. Integumentary potentials and wound healing. In R. B. Borgens; K. R. Robinson; J. W. Vanable, Jr., and M. E. McGinnis, *Electric Fields in Vertebrate Repair: Natural and Applied Voltages in Vertebrate Regeneration and Wound Healing.* Liss, New York, pp. 171–224.

Viale, G. 1916. Le correnti di riposo nel nervi durante la degenerazione e la rigenerazione. *Arch. di Fisiol.*, 14: 113–46.

1921. Ricerche elettrofisiologiche. I. La polarità delle correnti elettriche consecutive alle lesioni nei girini. *Arch. di Fisiol.*, 19: 243–7.

Waller, A. D. 1901a. On skin currents. I. The frog's skin. *Proc. R. Soc. Lond.*, 68: 480–94.

1901b. On skin currents. II. Observations on cats. *Proc. R. Soc. Lond.*, 69: 171–88.

1902a. On the "blaze-currents" of the incubated hen's egg. *Proc. R. Soc. Lond.*, 71: 184–93.

1902b. On skin currents. III. The human skin. *Proc. R. Soc. Lond.*, 70: 374–91.

11

Origin of the blastema cells in epimorphic regeneration of urodele appendages: a history of ideas

RICHARD A. LIVERSAGE

Charles Bonnet (1720–93) was opposed to the freethinking that prevailed in most scientific groups in his day. His views on natural philosophy agreed so well with Calvin's that some of Bonnet's writings were banned by Roman Catholic censors. Perhaps the most widely recognized preformationist, Bonnet believed in the incapsulation theory, that is, the idea that "germs" from all progeny of a species originated in the first female of that species. An ovist, he maintained that the spermatozoa of the male were required only for the activation of an ovum. Unfortunately, early in his career Bonnet suffered from both deafness and limited vision (Hays, 1972, pp. 171–81). Lazzaro Spallanzani (1729–99), however, shared Bonnet's views on ovist preformation and took it upon himself to draw public attention to their speculations on this subject, beginning in 1765 (Nordenskiold, 1928, p. 230). Spallanzani was probably the greatest and most versatile biological experimentalist of the eighteeth century (Glass, 1968, p. 170). Of particular interest to biologists who study vertebrate development are Spallanzani's original studies on amphibian limb and tail regeneration, published in 1769. (See Chapter 5 of the present volume.)

Trembley (1744) pioneered the study of regeneration in hydra. (See Chapter 4 of the present volume.) Bonnet and others confirmed the discovery but did not claim that hydra fragments contain complete minute hydra; rather, they believed the fragments contain preorganized particles from which new hydra can develop (Meyer, 1939, p. 56). Bonnet also studied regeneration of salamander limbs, as well as regeneration of severed parts of earthworms that yield perfectly normal worms. He concluded that "germs" exist not only in the ovaries of the female but also in regenerating animals and that germs are, without doubt, distributed throughout the body (Nordenskiold, 1928, p. 230). He states that the earthworm, like all other ani-

179

mals, must possess a soul and that the soul is one and indivisible. Because the earthworm regenerates, however, Bonnet was convinced that germs, which possess soul rudiments, must lie in all parts of the body, thereby allowing every regenerate to possess a soul (Nordenskiold, p. 231). It should be pointed out that Bonnet's and Spallanzani's "germs" existed in their theoretical thinking only in order to explain their deep convictions, as ovum preformationists, concerning the regulation of development of embryos and regenerating structures.

According to Morgan (1901, p. 23):

At present there are known two general ways in which regeneration may take place, although the two processes are not sharply separated, and may even appear combined in the same form. In order to distinguish broadly these two modes I propose to call those cases of regeneration in which a proliferation of material precedes the development of the new part, "epimorphosis." The other mode, in which a part is transformed directly into a new organism, or part of an organism without proliferation at the cut surface, [is] "morphallaxis."

(See Chapter 9 of the present volume.)

Weiss (1939) considered regeneration as the repair, by growth and differentiation, of damage suffered by an organism past the phase of primordial development. He thought that regenerative processes are fundamentally of the same nature and follow the same principles as the ontogenetic processes. He also states that regenerative capacity tends to vary inversely as the scale of organization. Generally speaking, the percentage of good regenerators is lower among the higher forms than among the lower, more simply organized forms. There is, however, no single animal group whose regenerative capacity could be portended from its position on the evolutionary scale. Weiss points out that some lower forms – for example, the ctenophores and the rotifers – have hardly any regenerative capability, whereas higher forms such as crustaceans and urodele amphibians possess considerable regenerative potential (p. 459).

For an appendage to regenerate, a stimulus is required. There must also be cellular components with the potential to give rise to new structures to replace the missing ones. According to Weiss, there also must be "organizing factors to make the right differentiations appear in the right places, to cause the proper alignments, movements and functional transformations of the cells, to direct the growth of the reconstituted parts into the proper proportions, and to control the resumption of functional activity in joint cooperation with the total

Figure 11.1. Oscar Emile Schotté (1895–1988), c. 1923–8, in Geneva, Switzerland. (Courtesy Amherst College, Amherst, Mass.)

parts. And, finally, there must be an adequate supply of food and other vital necessities" (p. 461).

In the early decades of the twentieth century, two investigators in particular made significant contributions to our understanding of tissue interactions in developing and regenerating systems of amphibians, concentrating on contributions to epimorphic limb regeneration in larval and adult urodeles (salamanders) (e.g., *Triton; Amb(l)ystoma* sp.; *Notophthalmus viridescens*, the adult newt).

The first of these contributions was made by Oscar E. Schotté (Figure 11.1). He made an extensive experimental study concerning the influence of nerves in urodele regeneration, the first of its kind since Todd's (1823) initial experiments demonstrating the involvement of nerves in the regeneration of limbs in salamanders. The results were published in 1926 from Schotté's doctor of science dissertation, completed in 1925 at the University of Geneva under Emile Guyénot (1885–1963). Schotté's original experiments, described in his thesis, provided the basis of the research he pursued throughout his tenure as professor of biology at Amherst College in Massachusetts (1934–65).

In 1928, Schotté was awarded a Rockefeller Foundation fellowship, following which he received a research appointment in the laboratory of Hans Spemann (1869–1941) at the University of Freiburg, in Germany. Upon joining Spemann's group, Schotté became aware of Spemann's obsession with the problem of tissue interactions in embryonic development. And this jaunty, highly motivated young scientist was to be instrumental in shedding light on another of Spemann's preoccupations – the role of chromosomes in development. The answer to *die Fragestellung* (placing the question) by Spemann emerged from Schotté's xenoplastic transplantation experiments. That is, Schotté raised salamander embryos to the larval stage, having transplanted ventral ectoderm of the frog gastrula to the mouth region of the salamander gastrula, and obtained salamander larvae with frog horny jaws and sucker. Furthermore, when he made reciprocal transplants of ventral ectoderm from the salamander gastrula to the mouth region of the frog gastrula, he obtained a pair of salamander balancing rods in frog tadpoles, as well as dentine teeth (Spemann and Schotté, 1932; Spemann, 1938, as cited in Hamburger, 1988, p. 42). To quote Victor Hamburger (1988), a former doctoral candidate of Spemann's: "This was undoubtedly one of the most spectacular experiments in experimental embryology. Considering that the two orders of amphibians, anurans [frogs] and urodeles [salamanders], have split-off from a common ancestor in the Paleozoic, 350 million years ago"(p. 42). Accordingly, anurans are more closely aligned with the main stem of evolution that gave rise to reptiles, birds, and mammals, than they are with urodeles.

As a former graduate student of Professor Schotté's, I can recall attending morning coffee break and afternoon "wine hour" in the laboratory, as well as spending any number of evenings at scientific meetings during which "the Professor" would engage colleagues and graduate students in rousing discussions of a wide variety of controversial subjects. This inevitably led to his retelling stories of his younger days in Europe, especially cossack tales and descriptions of his experiences as a preuniversity student at the gymnasia in Lovicz, Poland, and later in St. Petersburg, Russia. The story-telling would go on with his colorful adventures as a graduate student and instructor in Geneva (1920–8) under Guyénot, his postdoctoral experiences in Freiburg with Spemann, and his association with Ross G. Harrison (1932–4) at Yale University. Schotté idolized Harrison (1870–1959), always referring to him as "a fine gentleman, an 'apsolutely funtastik' embryologist whose pioneering research provided the basis for future experimental research pertaining to the nervous system, and the true father [developer] of tissue culture." Schotté was a dynamic regenerationist and lived a full, highly active professional life until his death in Amherst, on April 12, 1988, at the age of ninety-two (Liversage, 1978).

Third, and of particular importance, are the researches of Elmer G. Butler of Princeton University (Figure 11.2). As a young regenerationist, he spent a sabbatical year (1931) in Spemann's laboratory. There he met Schotté, and they became close colleagues as well as personal friends. As a result of Butler's initial investigations in Freiburg and later at Princeton, the Butler School of Dedifferentiation evolved (discussed later in this chapter). Butler's publications furnished fresh thinking and new findings concerning the origin of blastema cells in the regenerating salamander forelimb. After more than fifty years of scholarly pursuits, death came to this gifted scientist in Princeton, New Jersey, on February 23, 1972, at the age of seventy-two (Liversage, 1974).

Epimorphic and tissue regeneration

Among the vertebrates, exceptional ability to regenerate appendages is found in a number of larval and adult salamanders. Some of the best examples occur in the urodela order of amphibians, namely, in the families Ambystomatidae, Salamandridae, and Plethodontidae. Wholly aquatic salamanders, however, exhibit variability in limb regenerative potential. Although the capacity to regenerate limbs is severely restricted among most adult anurans, we find a few exceptions, such as *Xenopus laevis*, the South African clawed frog, which forms a heteromorphic (incomplete) spikelike regenerate.

There are two types of reparative regeneration among vertebrates involving cell proliferation, namely, epimorphic regeneration and tissue regeneration. Epimorphosis, as found in larval *Ambystoma* and the adult newt, requires the formation of an apical epithelial cap (AEC), dedifferentiation of the distal stump tissues, and the accumulation of a critical cell mass called the "blastema." Once these prerequisites are met, normal differentiation of the lost part of a limb ensues, and the regenerate becomes completely functional in a manner recapitulating embryonic development.

It appears that tissue regeneration of an appendage, such as is found in *X. laevis* adults, does not require the aforementioned prerequisites for differentiation; rather, after injury there is an immediate attempt to reconstitute some tissues, particularly the skin, cartilage, and bone. Accordingly, only incomplete (deficient) development of the regenerative outgrowth ensues.

Figure 11.2. Elmer Grimshaw Butler (1900–72), c. 1955–60, in Princeton, New Jersey. (Courtesy of Princeton University)

The following *specific requirements* and *regulatory factors* are essential for epimorphic regeneration of an amputated appendage in the adult newt: First, *amputational injury* is the initial stimulus for limb and tail regeneration. This is followed by *epithelial wound healing*, or wound closure. *Wound repair* ensues, in which cellular debris is removed from the injury site. Subsequently a thickened apical epithelial cap, free of underlying dermis, covers the wound. *A cell source* is required for blastema formation; redifferentiation is delayed until a "critical mass" of blastema cells has formed. *Adequate innervation* is vital in support of cell proliferation (e.g., blastema growth) and, to some degree, for morphogenesis. A *hormonal milieu* (Liversage, McLaughlin, and McLaughlin, 1985) is necessary for cell growth, differentiation, and metabolism. And, a definite organization of cells in the stump and regenerate is required, in order to provide for the transfer of *morphogenetic information* between and among cells, which is then processed by encoded positional memory. (See Chapter 12 in the present volume.) *Bioelectric fields* are essential, in the early stages of amphibian appendage regeneration. (See Chapter 10 in the present volume.) Last, analogies between the regeneration blastema and the development of tumors suggest that the *immune system* may influence regeneration in vertebrates (Sicard, 1985). For further consideration of epimorphic regeneration in the adult newt forelimb, see the reviews by Goss (1969), Schmidt (1968), Wallace (1981), Sicard (1985), and Liversage (1987).

Origin of the blastema

Spallanzani first described and sketched urodele regeneration blastemas in the late 1760s, and, under Bonnet's influence, wondered if they were simply the product of expanding "germs." From the latter part of the nineteenth century to the present, theories have been put forth and experimental and analytical approaches undertaken to determine the origin of the blastema cells during epimorphic regeneration in urodele appendages. As a result, "Four Schools of Interpretation" should be taken into account – those espousing namely: (1) the blood cell origin (Colucci, 1884; Hellmich, 1930, 1931); (2) the reserve cell source (Weiss, 1939; see also Cameron, Hilgers, and Hinterberger, 1986); (3) the wound epithelial or epidermal cell origin (Godlewski, 1928; S. M. Rose, 1948); and (4) the dedifferentiation of mesodermal stump tissues source (Butler, 1933, 1935; Butler and O'Brien, 1942; Chalkley, 1954; Hay and Fischman, 1961; and other, more recent work).

Blood origin

Early investigators described the cells in the young regenerative limb bud in amphibians (salamanders) as embryonic. Colucci (1884), an Italian Swiss, first proposed that blastema cells originated in white blood cells and other tissues, in varying amounts. "Following amputation of the limb of a newt, in both the juvenile and adult, by a process of new formation a bud forms comprised of and originating from the white blood cells emigrating from vessels surrounding the wound and from the marrow of the bone or cartilage of the juvenile." In the adult, Colucci also describes the "breaking down of the bone of the femur into spongy bone [*osseo spugnoso*] and defacement of the epiphysis, becoming a transformed cartilage surrounding the marrow and having an embryonic character. Following the progression of the cartilage tissues created by proliferation and differentiation, the bone is ossified and vascularized." Colucci concludes that "regeneration of limb and tail in the newt occurs by means of the granulation and differentiation of the various constituted tissues and in large degree by emigrating leucocytes. These are persistent embryonic elements because of their incessant formation and [they] are destined to meet the continuous need of the organism, both in a normal and pathologic state" (pp. 554–6).

The modern concept of cells and tissues, such as Schleiden and Schwann's cell theory, was essential to the development of contemporary theories as to the cell source. Hellmich (1930) described two major classes of cells in the adult urodele regenerate: hematogenetic cells and histiogenetic, or blood and mesenchymal (embryonal), cells. In 1931, Hellmich reported that in the older regeneration bud these various cells begin to have similar structure; he designated them "regeneration cells." Both Hellmich and Colucci emphasized that the hematogenetic cells make a greater contribution to the "regeneration cell" population than do the "various constituted tissues" (Colucci, pp. 554–6).

Using adult *Plethodon cinereus* and *Eurycea bislineata*, Hellmich (1931) showed that the process of regeneration is very similar to that found in *Amblystoma mexicanum*. That is to say, in the early regeneration blastema beneath the (wound) "epithelium of undifferentiated ectoderm cells" (p. 304), he identified cells of hematogenetic origin. "There is an extremely high number of a special type of round cell, readily distinguishable by their light cytoplasm and polymorphic nucleus – the special leucocytes." Associated macro- and micro-lymphocytes, mostly with a round nucleus and a small rim of cytoplasm were present, as well as a small number of eosinophile(s). "Also, erthyrocytes, partly shrunken and necrotic, are found in the regenerating blood vessels [and these vessels are present] between the different cells [in the blastema] and beneath the ectoderm" (p. 304). According to Hellmich, histiogenetic cells could also be observed, especially in the later regenerative stages of these adult urodeles. Hellmich considered these cells to be similar to "common fibroblasts" (p. 304) from which he thought they might be derived. He describes the larger of these cell types as being like the undifferentiated mesenchymal cells that had been described in *Amblystoma*. He also observed osteoblasts within the interior of the bud, cells that had been described, incorrectly, by earlier investigators as giant cells. He assumed that osteoblasts gradually lose the bony matrix in which they are embedded during the regenerative process. This comment suggests the concept of dedifferentiation of the distal stump tissues. In his 1931 publication, Hellmich stated that, after complete extirpation of the larval limb bud in urodeles, regeneration does not ensue, because all cells of *die Anlage* (the plan or design) and its immediate environment have been removed. Therefore, he concluded that mesenchymal cells also make a contribution to the regenerative process; fibroblasts are "transformed" (Morgan, 1901, p. 23) into the cells of the blastema along with blood cells, particularly macrophages and lymphocytes as well as nucleated red blood corpuscles.

Reserve cell source

The early investigators of epimorphic regeneration in amphibian appendages presumed that each stump tissue generated more of the same cells following amputation. But it was shown that when the skeletal elements in an adult newt limb are removed, new bone will develop perfectly in the regenerate, distal to the amputation level, from the cells of the blastema (Fritsch, 1911; Weiss, 1925). In 1939, Weiss concluded that the blastema cells arise in the proximity of the wound. He postulated that the blastema forms owing to the presence in the limb of undifferentiated "reserve cells," of connective tissue origin, which do not actually differentiate and thereby reserve the ability (competence) to differentiate into a variety of mesodermal limb tissues when the need arises. He theorized that these reserve cells give rise to new bone tissue in a limb deboned before amputation. He postulated that during this period differentiation resulted from the separation of cells, hence differentiation is not reversible. Indeed, once a cell has assumed a definite course of differentiation, it is not able to change to a different course or different cell type (Weiss, 1939, p. 468). Subsequent studies indicate that reserve cell populations, may, in some cases, be a factor in regenerative processes.

Hay and Doyle (1973) state that skeletal muscle in adult metamorphosed urodeles regenerates in the absence of satellite cells. These satellite cells are undifferentiated, spindle-shaped, flattened, mononucleate cells, of no apparent function, lying between the basement membrane and the sarcolemma of a muscle fiber (Mauro, 1961). But Cameron, Hilgers, and Hinterberger (1986), in experiments on adult newt muscle, showed that postsatellite cells, a type of myogenic reserve cells (Popiela, 1976), give rise to regenerated myotubes in vitro. They concluded that the presence of these "myogenic postsatellite cells in adult newt muscle provides an alternative to the hypothesis that myoblasts [muscle progenitor cells] in regenerating adult urodele limbs arise through dedifferentiation of mature myofibres" (p. 607).

Epithelial cell origin

Godlewski (1928) was the first to propose that wound epithelium contributed cells to the mesenchymal blastema. Godlewski and S. M. Rose (1948) stressed the close contact between the apical epithelium and the blastema; they noted that tonguelike projections composed of inner basal epithelial cells projected into the blastema cell mass. Rose maintained that these cells actually underwent metaplasia, entered the blastema, and became blastema cells. Other investigators have also observed these prominent projections from the apical epithelium (Scheuing and Singer, 1957; Schmidt, 1958). Godlewski too noted that the epithelium covering the wound surface was different from the epidermis at the lateral edges of the wound. On the basis of cell counts, Rose concluded that there was a considerable loss of epithelial cells covering the wound concomitant with an increase in the blastema cell population.

Rose, Quastler, and Rose (1955) stimulated regeneration when lead-shielded, normal, proximal epidermis was observed to migrate and to cover the distal, x-irradiated part of an adult newt limb from which the x-irradiated skin covering had been removed. X-irradiation of cells inhibits the formation of the mitotic spindle and thereby interferes with cell division. Rose and Rose (1965) injected adult newts with tritiated thymidine, a radioisotope-labeled DNA-component molecule, the day before amputation and then stripped the limb skin to a point above the elbow and amputated near the wrist. The amputation surface was covered by a labeled epithelial cell layer that migrated distally. They concluded that these limbs regenerated normally, inasmuch as labeled cells were present within the blastemal mesenchyme together with unlabeled cells. When this labeling experiment was repeated and labeled skin was substituted for unlabeled skin, limb regeneration ensued, and they state that nonlabeled cells were found in the blastema. Again, Rose and Rose concluded that metaplasia, or transmutation, occurred and that epithelial cells became blastemal mesenchyme.

Elizabeth Hay (1952) transplanted epidermal cells with three sets of chromosomes to regenerating hind limbs of *Rana pipiens* tadpoles (in which the cells have only two sets of chromosomes), in order to determine whether epithelial metaplasia occurred. But very few of her cases showed the presence of, and thereby the possible transformation of basal epithelial cells into, embryonic-like or mesenchymal-like blastema cells. Moreover, Bodemer (1958) noted that charcoal particles placed under the epithelium were removed by cells of the inner projecting tongues of epithelium. The epithelial cells apparently engulfed these particles, and the cells were later exteriorized. Therefore, the wound epithelium was not contributing cells to the blastemal mesenchyme; instead, cells were moving distally, in the opposite direction, into the external environment. (see also Salpeter and Singer, 1960; Singer and Salpeter, 1961.)

In addition to the observations of Hay (1952) and Bodemer (1958), considerable published experimental and analytical data now support and, thereby, dispute direct cellular contributions by the apical wound epithelium, which originates from the more lateral epidermis, to the development of the blastema. Despite the aforementioned cell–cell interaction, however, contact between the epithelium and the mesenchymal cells is essential for the development of the blastema. Prevention of this contact inhibits regeneration. That is to say, when the amputation surface of an appendage is overlain by whole skin, regeneration ceases because of the presence of the intervening dermal layer.

Dedifferentiation of mesodermal stump tissues as the cell source

Driesch (1902) and later Schultz (1907) first used the term "dedifferentiation" in experiments concerning the regeneration of the branchial chamber in *Clavellina* (Ascidian). Their interpretation was that differentiated cells reverted to an embryonic condition; hence the term "dedifferentiation" was used to describe this phenomenon by which the cells again became pluripotent. These cells then combined with the recuperating branchial region as nonspecialized cells. Schaxel (1914) disputed the capability of cells to dedifferentiate and held that once development was accomplished it was not reversible. He considered "dedifferentiated cells" simply as dying cells. Schaxel and others surmised that the regenerate was composed of "reserve cells" which were found throughout an organism's body. After injury, these cells proliferated and provided the source of cells for regeneration. Accordingly, Weiss (1939) argued that the term dedifferentiation should be discarded in favor of "modulation."

The observations and experimental findings of Hellmich (1930, 1931), Rose (1948), Hay (1952), and others, as well as the proposal by Rose and Godlewski (1928) of a significant epithelial, metaplasial contribution of cells to the blastemal mesenchyme were placed in serious question by the incisive observations and classical experiments of Elmer G. Butler. His initial histological observations and subsequent x-irradiation experiments, conducted in Freiburg and Princeton, provide evidence that the blastema cells of the regenerating urodele limb arise from "local" mesodermal tissues immediately proximal to the amputated limb site. And, when "E.G.B." described these observations in his 1933 and 1935 publications, he used the controversial term "dedifferentiation" anew, a term all too familiar to those of us who were members of Butler's "Fourth Floor Family." This latter expression was his way of showing parental affection for his doctoral students and postdoctoral fellows, who spent many enlightening hours with this distinguished scientist and gentleman on the top level of Guyot Hall at Princeton.

In a classical experiment by Butler and O'Brien (1942), an entire salamander larva (*Eurycea bislineata*), except for the left knee, was lead-shielded and the knee was exposed to x-irradiation (2.5–5.0 kilorads). Amputation of the shielded ankle resulted in regeneration. A subsequent limb amputation through the knee region, which had been exposed to x-irradiation, showed that a 5-kilorad irradiation blocked regeneration. A third amputation through the lead-protected thigh of the same animal gave normal limb regeneration, demonstrating that the cells for the regeneration blastema had to be of "local origin." The authors concluded that cells dedifferentiated from the mesodermal cut stump tissues (Figure 11.3).

The results of Chalkley's studies (1954, 1959) of normal forelimb regeneration in *N. viridescens,* initially conducted in Butler's laboratory, differ significantly from Godlewski (1928) and Rose's interpretation (1948) of a wound epithelial cell origin. Chalkley states that "one of the conspicious consequences of limb amputation is the initiation of tissue dissociation and cellular dedifferentiation in subepidermal (mesodermal) tissues at the amputation level." David (1934) observed that mitosis ensues as cells dedifferentiate. He stated that the proliferation continues among these embryoniclike cells, and they accumulate beneath the apical epithelial cap. Chalkley concluded that the periosteum, muscle, and the different layers of connective tissue (i.e., mesodermal tissues) appear to provide the blastemal cells. Indeed, his cell counts and mitotic indices show that approximately 85 percent of the cells in the developing blastema come from the connective tissues. He also concluded that the wound epithelium does not contribute appreciable numbers of cells to the mesenchymal mass of the blastema. Manner (1953) also concluded, after making mitotic cell counts in *N. viridescens* forelimb blastemas, that cells originate from the distalmost cut stump tissues. Although Chalkley quite clearly demonstrated that the regeneration blastema cells originate as a result of the phenomenon of mesodermal stump tissue dedifferentiation, his studies do not explain, nor does he tend to speculate as to, the mechanism(s) involved in this process.

Hay (1959) observed the separation of muscle syncytia into mononucleate cells, the anucleated portions of which were lysed during the dedifferentiation process. The nucleated, dissociated cells became oriented toward, and appeared to move in the direction of, the developing blastema. During this phase, the parallel organization of the myofibrils was lost, whereupon the fibrils became fragmented and disappeared. The presence of ribonuclear granules and large nucleoli were indicators of increased protein synthesis. The former muscle cells could not be identified by their form, but their high synthetic activity was indicative of a cell that was not degenerating. Upon further observation, Hay (1974) described the stages through which striated muscle cells pass to become embryoniclike or mesenchymal cells. Many vesicles form in the cytoplasm, and the mitochondria move to one end of the cell, adjacent to the nucleus. While enclosing a small region of the inner cytoplasm and the nucleus, the vesicles align; they then fuse with the outer membrane, thus freeing the nucleus and a small amount of cytoplasm from the rest of the still-intact muscle mass. Aggregations of such cells form the source of progenitor, or "seed," cells (Liversage, McLaughlin, and McLaughlin, 1985), which, in turn, proliferate under the influence of nerves, the apical epithelial cap, and the hormonal milieu (ibid.) and result in the formation of a blastema. (See also Globus, 1978, on "tripartite control of the cell cycle," pp. 863–6.)

Some persuasive results arguing against an epithelial origin of the regeneration blastema cells are from investigations using tritiated thymidine (a radionucleotide) by Riddiford (1960), Hay and Fischman

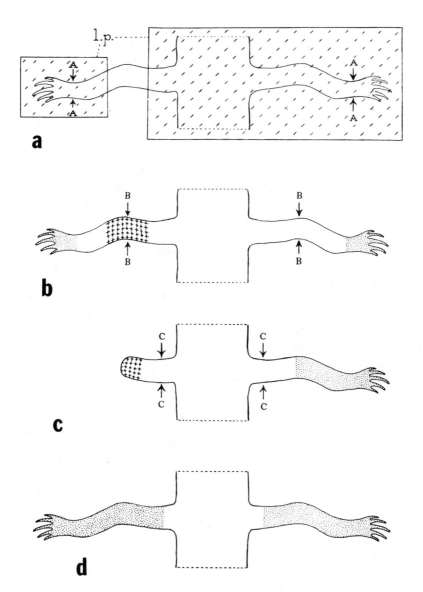

Figure 11.3. (*a*) Diagram of arrangement for localized radiation. The entire larva was shielded with lead plate (*l.p.*), with the exception of the region of the knee joint of the left hind leg. Arrows (*A–A*) indicate level of first limb amputation. (*b*) Same larva 3 weeks after irradiation and first limb amputation. Stippled areas represent regenerated structures; crosshatching indicates region originally irradiated; arrows (*B–B*) show level of second amputation. (*c*) Larva 9 weeks after second limb amputation. Stippled area represents regenerated structures; crosshatching indicates nonre-

(1961), O'Steen and Walker (1961), and Namenwirth (1974). For example, Riddiford radiolabeled epidermal transplants, which then formed the epithelium covering the unlabeled stump tissues of limbs. She found that the label remained in the epithelial transplant; that is to say, no labeled epithelial cells were found among the blastema cell population. When the reciprocal experiment was performed, whereby tritiated thymidine – labeled stump tissues were covered by unlabeled epidermal transplants, the label was found in the wound epithelium. This was due, however, to the movement of cell remnants and exteriorized cell waste products, as well as macrophages, from the mesenchymal blastema into and through the wound epithelium and from there outward into the external environment (Salpeter and Singer, 1960; Singer and Salpeter, 1961).

In the classical experiments of Hay and Fischman (1961), autoradiographic methods were employed to determine whether tritiated thymidine was incorporated by cells of normally regenerating adult newt limbs during the preblastemic, blastemic, and early postblastemic stages. In this manner they followed the migration of cells labeled by the radioisotope. Results showing areas of DNA synthesis during blastema formation, and therefore the role of the inner mesodermal tissues and epithelial cells, were obtained. In series 1, regenerating limbs 1–28 days after amputation were fixed on the day the newts were injected with isotope. The results show that DNA synthesis begins 4–5 days after amputation in dedifferentiating mesodermal stump tissues, about 1 millimeter proximal to the amputation level. The authors observed that cells of the stump tissues synthesized DNA at increased rates between days 10–20 after amputation. But the epithelium that migrates over and covers the wound discontinues DNA synthesis by 2 days after amputation, unlike the epithelium proximal to the amputation surface. These latter cells show greatly increased DNA synthesis by 8 days postamputation. By 10–15 days, the migrated cells increase the thickness of the covering epithelium. After blastema formation, however, the regenerate begins to elongate, and the distal apical epithelial cap thins out, whereupon the remaining apical epithelial cells begin DNA synthesis again.

In series 2, the limbs were treated with tritiated thymidine in the earlier stages of regeneration, at 5, 10, and 15 days postamputation.

Caption to Figure 11.3 *(cont.)*
generating irradiated region; arrows (*C–C*) show level of third amputation. (*d*) Same larva 4 weeks after third amputation. Stippled areas indicate regenerated limbs. (Reprinted from Butler and O'Brien, 1942, *Anat. Rec.* 84: 407–13, Figures 1–4, by permission of Alan R. Liss Publishers, New York, USA.)

Subsequently, limbs were fixed daily for up to 5 days after treatment. The dedifferentiating mesodermal cells incorporated tritiated thymidine on the day of injection, but the apical epithelial cells did not. The label subsequently was detected in blastema cells. Hay and Fischman concluded that these blastema cells were derived from the dedifferentiated mesodermal stump tissues.

In series 3, newts were injected with tritiated thymidine prior to forelimb amputation. The epidermis incorporated the label, but the internal stump tissues did not. After amputation, labeled epidermal cells migrated over the wound surface. An apical epithelial cap developed and remained labeled during blastema formation, but the mesenchymelike blastema cells were not labeled. Although radiolabeled, the apical epithelial cap cells did not exhibit DNA synthesis nor the characteristics of dedifferentiating cells, and no transformation of epithelial cells into blastema cells could be demonstrated.

Mettetal (1939) regards the connective tissue proper as the sole source of blastema cells. Schmidt (1968) states that fibroblasts may be the exclusive source of cells in the adult newt blastema. As well, Dunis and Namenwirth (1977) and Muneoka, Fox, and Bryant (1986) consider that fibroblasts from the dermis contribute significantly to the blastema. Maden (1977), however, proposes that under certain experimental conditions ("paradoxical" regeneration), the axolotl (salamander) blastema can be entirely derived from Schwann cells. Wallace (1981) suggests that the predominance of connective tissue mitosis (85 percent, as observed by Chalkley, 1954, during cell accumulation and growth of the blastema) reflects the ubiquity of the fibroblasts. Indeed, all mesodermal tissues of the distal stump are involved, to a greater or lesser extent, in the contribution of cells to the adult newt blastema by the process of dedifferentiation. It is clear that we do not yet fully understand the dynamics of blastema formation, and we continue to share the challenge that it presented to our intellectual forebearers (e.g., Spallanzani and Morgan).

These "Four Schools of Interpretation" as to the source of the progenitor, or seed, cells in epimorphic regeneration of urodele limbs all have validity. That is to say: (1) The studies of Butler, in collaboration with members of his research group at Princeton – including observations by Chalkley, (1954, 1959), the experiments of Hay and Fischman (1961), and the current investigations of Brockes and his group (Kintner and Brockes, 1984; Casimer, Gates, Patient, and Brockes, 1988), to mention only a few – have conclusively demonstrated that dedifferentiation of the mesodermal stump tissues in regenerating larval and adult urodele forelimbs is the main, and probably the only, provider of progenitor cells for the accumulation blastema. (2) Although the hematogenic cells (Hellmich, 1930, 1931; Colucci, 1884) are not the source of mesenchymal blastema cells, they provide the blastema cells with gas exchange and also lymphocytes and macrophages, as well as osteoclasts (Fischman and Hay, 1962), which are involved in the wound-repair stage and in bone and cartilage matrix breakdown, respectively. (3) Weiss's "reserve cell" interpretation postulates that the main source of cells is fibroblasts from the neurolemma and muscle sheaths. Certainly these fibroblasts represent one of the blastema cell sources, as demonstrated by Chalkley (1954). Cameron and her colleagues (1986) have demonstrated the presence of "postsatellite cells," which they hypothesize are a type of myogenic "reserve cell" that gives rise to skeletal muscle in the adult newt regenerate. (4) Godlewski and Rose's assumption that the epithelium ("epidermis") is the main source of blastema cells has been clearly demonstrated to be incorrect, at least in adult newts, by a number of investigations. Nevertheless, the epithelial cells provide covering for the fresh wound, and subsequently a multilayered apical epithelial cap forms. In addition, there is evidence that the overlying apical cap cells may provide proteolytic or histolytic enzymes that assist in the dedifferentiation process as a result of their cell–cell contact and interaction with the remaining distal stump tissues (Grillo, LaPiere, Dresden, and Gross, 1968; Schmidt, 1968; and others). Presumably, these enzymes break down collagen fibrils and tissue matrices during the initial stages of regeneration, thus freeing cells that become progenitors of the developing blastema. The apical epithelial cap, present at the accumulation blastema stage and subsequently, also keeps the dedifferentiated seed cells and their progeny in the cell cycle. Presumably, the apical cap provides a "division signal" (Globus, 1978) that acts in concert with the innervation of the blastema in the presence of insulin (as well as the hormonal milieu, Liversage et al., 1985).

The theoretical considerations of E. G. Butler, coupled with his published observations and experimental findings, which purport that "dedifferentiation" is the source of progenitor cells of the blastema in epimorphic regeneration, have stood the test of time. His interpretations have never been contradicted; rather, they have been accepted as the foundation to which today's developmental, cellular, and molecular regenerationists have added further experimental and analytical proof. Butler's classical studies, as well as those of Donald Chalkley, Elizabeth Hay, Donald Fischman, and other more recent investigators, have brought us closer to answering the question of why some vertebrates such as urodele amphibians are capable of regenerating completely normal appendages whereas the human primate cannot.

ACKNOWLEDGMENTS

I wish to express my appreciation to Donna R. Wheeler and Rossana Soo for their expert word-processing skills and proofreading of the manuscript; to Professor Domenico Pietropaola of the Department of Italian Studies and Roy D. Pearson of the Science and Medicine Library, University of Toronto, for translating the Colucci (1884) publication; and to the staffs of the Museum of Comparative Zoology and Widner Libraries of Harvard University for their cooperation in providing most of the older literature. Funds supporting the writing of this chapter were provided for me by NSERC of Canada grant A-1208.

REFERENCES

Bodemer, C. W. 1958. The development of nerve-induced supernumerary limbs in the adult newt, *Triturus viridescens*. *J. Morphol.* 102: 555–81.

Butler, E. G. 1933. The effects of x-radiation on the regeneration of the forelimb of *Amblystoma* larvae. *J. Exp. Zool.* 65: 271–316.

1935. Studies on limb regeneration in X-rayed *Amblystoma* larvae. *Anat. Rec.* 62: 95–307.

Butler, E. G., and J. P. O'Brien. 1942. Effects of localized X-radiation on regeneration of the urodele limb. *Anat. Rec.* 84: 407–13.

Cameron, J. A.; A. R. Hilgers; and T. J. Hinterberger. 1986. Evidence that reserve cells are a source of regenerated adult newt muscle in vitro. *Nature, Lond.* 321: 607–10.

Casimer, C. M.; P. B. Gates; R. K. Patient; and J. P. Brockes. 1988. Evidence for dedifferentiation and metaplasia in amphibian limb regeneration from inheritance of DNA methylation. *Devel.* 104: 657–68.

Chalkley, D. T. 1954. A quantitative histological analysis of forelimb regeneration in *Triturus viridescens*. *J. Morphol.* 94: 21–70.

1956. The cellular basis of limb regeneration. In *Regeneration in Vertebrates*, ed. C. S. Thornton. University of Chicago Press, Chicago, pp. 34–58, 102–4.

Colucci, V. 1884. Intorno alla rigenerazione degli arti es della coda nei *Tritoni*. Studio sperimentale. [*Mem. Ric. Accad. Sci Inst. Bologna*], ser. 4, 6: 501–66.

David, L. 1934. La contribution du matériel cartilagineux et osseaux au blastème de régénération des membres chez les amphibiens urodeles. *Arch. d'Anat. Microsc.* 30: 217–34.

Driesch, H. 1902. Studien über das Regulations vermögen der Organismen. 6. Die Restitution der *Clavellina lepadiformis*. *Arch. EntwMech.* 14: 247.

Dunis, D. A., and M. Namenwirth. 1977. The role of grafted skin in the regeneration of x-irradiated axolotl limbs. *Dev. Biol.* 56: 97–109.

Fischman, D. A., and E. D. Hay. 1962. Origin of osteoclasts from mononuclear leucocytes in regenerating newt limbs. *Anat. Rec.* 143: 329–38.

Fritsch, C. 1911. Experimentelle studien über Regenerationsvorganger des Gleidmassenskeletts. *Zool. Jahrb. Abt. Physiol.* 30: 377–472.

Glass, B. 1968. Heredity and variation in the eighteenth century concept of the species. In *Forerunners of Darwin, 1745–1859*, ed. B. Glass, O. Temkin, and W. L. Straus. John Hopkins University Press, Baltimore, pp. 42–8, 170.

Globus, M. 1978. Neurotrophic contribution to a proposed tripartite control of the mitotic cycle in the regeneration blastema of the newt (*Notophthalmus [Triturus] viridescens*). *Am. Zool.* 18: 855–68.

Godlewski, E. 1928. Untersuchungen über Auslösung und Hemmung der Regeneration beim Axolotl. *Arch. f. EntwMech.* 114: 108–43.

Goss, R. J. 1969. *Principles of Regeneration.* Academic Press, New York.

Grillo, H. C.; C. M. LaPierre; H. H. Dresden; and J. Gross. 1968. Collagenolytic activity in regenerating forelimbs of the adult newt (*Triturus viridescens*). *Dev. Biol.* 17: 571–83.

Hamburger, V. 1988. *The Heritage of Experimental Embryology: Hans Spemann and the Organizers.* Oxford University Press, New York.

Hay, E. D. 1952. The role of epithelium in amphibian limb regeneration, studies by haploid and triploid transplants. *Am. J. Anat.* 91: 447–81.

    1959. Electron microscopic observation of muscle dedifferentiation in regeneration *Amblystoma* limbs. *Dev. Biol.* 1: 555–85.

    1974. Cellular basis of regeneration. In *Concepts of Development*, ed. J. Lash and J. R. Whitaker. Sinauer Assoc., Stamford, Conn., pp. 404–28.

Hay E. D., and C. M. Doyle. 1973. Absence of reserve cells (satellite cells) in nonregenerating muscle of mature newt limbs. *Anat. Rec.* 175: 339–40.

Hay E. D., and D. A. Fischman. 1961. Origin of the blastema in regeneration limbs of the newt *Triturus viridescens:* An autoradiographic study using tritiated thymidine to follow cell proliferation and migration. *Dev. Biol.* 3: 26–59.

Hays, H. R. 1972. *Birds, Beasts and Man.* Penguin, Baltimore, pp. 171–81.

Hellmich, W. 1930. Untersuchungen über Herkunft und Determination des regenerativen Materiels bei Amphibien. *Arch. f. EntwMech.* 121: 135–202.

    1931. Histology of regeneration in different species of adult and larval urodeles. *Anat. Rec.* 48: 303–7.

Kintner, C. R., and J. P. Brockes. 1984. Monoclonal antibodies identify blastemal cells derived from dedifferentiating muscle in newt limb regeneration. *Nature* 308: 67–9.

Liversage, R. A. 1974. In Memoriam: Elmer Grimshaw Butler. *J. Exp. Zool.* 190: 129–32.

    1978. Dedication to Oscar E. Schotté. *Am. Zool.* 18(4): 825–7.

    1987. Limb regeneration in vertebrates: Regulatory factors. *Biochem. Cell Biol.* 65: 726–9.

Liversage, R. A.; D. S. McLaughlin; and H. M. G. McLaughlin. 1985. In *Regulation of Vertebrate Limb Regeneration*, ed. R. E. Sicard. Oxford University Press, New York.

Maden, M. 1977. The regeneration of positional information in the amphibian limb. *J. Theor. Biol.* 69: 735–53.

Manner, H. W. 1953. The origin of the blastema and of new tissues in the regenerating forelimb of the adult *Triturus viridescens* (Rafinesque). *J. Exp. Zool.* 122: 229–57.

Mauro, A. 1961. Satellite cells of skeletal muscle fibres. *J. Biophys. Biochem. Cytol.* 9: 493–5.

Mettetal, C. 1939. La régénération des membres chez la salamandre et le triton. *Arch. Anat. Histol. Emb.* 28: 1–214.

Meyer, A. W. 1939. *Rise of Embryology*. Stanford University Press, Stanford, Calif., pp. 54–85.

Morgan, T. H. 1901. *Regeneration*. Macmillan, London.

Muneoka, K., W. F. Fox, and S. V. Bryant. 1986. Cellular contribution from dermis and cartilage to the regenerating limb blastema in axolotl. *Dev. Biol.* 116: 260–5.

Namenwirth, M. 1974. The inheritance of cell differentiation during limb regeneration in the axolotl. *Dev. Biol.* 41: 42–56.

Nordenskiold, E. 1928. *History of Biology*. Tudor, New York, pp. 243–52.

O'Steen, W. K., and B. E. Walker. 1961. Radioautographic studies of regeneration in the common newt. II. Regeneration of the forelimb. *Anat. Rec.* 139: 547–56.

Popiela, H. 1976. In vivo limb tissue development in the absence of nerves: A quantitative study. *Exp. Neurol.* 53: 214–26.

Riddiford, L. M. 1960. Autoradiographic studies of tritiated thymidine infused into the blastema of the early regenerate in the adult newt, *Triturus. J. Exp. Zool.* 144: 25–32.

Rose, F. C.; H. Quastler; and S. M. Rose. 1955. Regeneration of x-rayed salamander limbs provided with normal epidermis. *Science* 122: 1018–19.

Rose, F. C., and S. M. Rose. 1965. The role of normal epidermis in recovery of regenerative ability in x-rayed limbs of *Triturus. Growth* 29: 361–93.

Rose, S. M. 1948. Epidermal dedifferentiation during blastema formation in regenerating limbs of *Triturus viridescens. J. Exp. Zool.* 108: 337–61.

Salpeter M. M., and M. Singer. 1960. Differentiation of the submicroscopic adepidermal membrane during limb regeneration in adult *Triturus*, including a note on the use of the term basement membrane. *Anat. Rec.* 136: 27–40.

Schaxel, J. 1914. *Verh. d. Deutschen Zool. Ges.* 24: 122.

Scheuing, M. R., and M. Singer. 1957. The effects of microquantities of beryllium ion on the regenerating forelimb of the adult newt, *Triturus. J. Exp. Zool.* 136: 301–27.

Schmidt, A. J. 1958. Forelimb regeneration of thyroidectomized adult newts. II. Histology. *J. Exp. Zool.* 139: 95–136.

    1968. *Cellular Biology of Vertebrate Regeneration and Repair*. University of Chicago Press, Chicago.

Schotté, O. E. 1926. Système nerveux et régénération chez le *Triton. Rev. Suisse Zool.* 33: 1–211.

Schultz, E. A. 1907. Über Reduktionen. III. Die Reduktion und Regeneration des abgeschnittenen Kiemenkorbes von *Clavellina lepadiformis. Arch. f. EntwMech.* 24: 503.

Sicard, R. E. 1985. Leukocyte and immunological influence on regeneration of amphibian forelimbs. In *Regulation of Vertebrate Limb Regeneration*, ed. R. E. Sicard. Oxford University Press, New York.

Singer, M., and M. Salpeter. 1961. Regeneration in vertebrates: The role of the wound epithelium. In *Growth in Living Systems*, ed. M. X. Zarrow. Basic Books, New York, pp. 277–311.

Spallanzani, L. 1769. *An Essay on Animal Reproductions*. (Translated from the Italian, 1768, by M. Maty.) T. Becket, London.

Spemann, H. 1938. *Embryonic Development and Induction*. Yale University Press, New Haven.

Spemann, H., and O. E. Schotté. 1932. Über xenoplastische Transplantation als Mittel zur Analyse der embryonalen Induktion. *Naturwissenschaften* 20: 463–7.

Todd, T. J. 1823. On the process of reproduction of the members of the aquatic salamander. *Q. J. Sci., Lit., Arts* 16: 84–96.

Trembley, A. 1744. *Mémoires, pour servir à l'histoire d'un genre de polypes d'eau douce*. Verbeek, Leiden.

Wallace, H. 1981. *Vertebrate Limb Regeneration*. Wiley, New York.

Weiss, P. 1925. Abhängigkeit der Regeneration entwickelter Amphibienextremitaten vom Nervensystem. *Arch. f. Mik. Anat.* 104: 317–58.

1939. *Principles of Development*. Holt, New York.

Yntema, C. L. 1959. Regeneration in sparsely innervated and aneurogenic forelimbs of *Amblystoma* larvae. *J. Exp. Zool.* 140: 101–23.

12

Morgan's ambivalence: a history of gradients and regeneration

LEWIS WOLPERT

In 1745, Charles Bonnet suggested that the regeneration of a head at one end of a worm and a tail at the other is due to fluids that flow forward and backward and act on "head germs" at the anterior end and on "tail germs" at the posterior end. Thus the accumulation of the head-stimulating fluid at the anterior end was supposed to "awaken" the germ of a head that lies at the anterior end. But "the assumption of head and tail-germs and also the forward flow of certain substances, and the posterior flow of other kinds of substances, are entirely ficti- tious assumptions, which from our modern point of view would be more difficult to account for than the phenomenon of regeneration itself." So wrote T. H. Morgan (1903). It was a sentiment to be echoed by some biologists in their attitude toward gradients for the next half-century. And as we shall see, Morgan himself, who invented the "gradient" concept, was ambivalent – almost schizophrenic – on this point.

Nineteenth-century background

Bonnet's ideas were not precisely about gradients in regeneration, but they ran along lines that were later adopted in relation to gradients, in that he proposed the unequal distribution of substances that stimulate regeneration and also implied that polarity is linked to the direction of flow of these substances. Perhaps, Morgan suggests (1901, p. 41), Bonnet had something like blood in mind but we might give him the benefit of the doubt and assume he meant the flow of a more "subtle" fluid.

Bonnet was a preformationist, and just as the egg was supposed to contain a preformed germ, so he thought regeneration resulted from the development of latent germs in the adult animal. The idea was that germs for the different parts that could be regenerated were distributed in the regions that could regenerate. This idea of prefor-

201

mation was very much part of Weismann's (1893) conception of regeneration a hundred and fifty years later. For regeneration in hydra, he, like Bonnet, assumed determinants for the proximal and distal ends and assumed that an individual cell must be capable of dividing in different planes and giving rise to different regions of the body. Weismann's views on regeneration conformed to his ideas on development. He conceived of ontogeny here as a series of gradual, qualitative changes in the nuclear substance – idioplasm – of the cells as the cells multiplied. The determinants of cell character, he thought, are contained within the idioplasm, and thus the idioplasm of the cells of the regenerating limb must contain the determinants of all succeeding bone cells. He assumed that in regeneration each specific cell could reproduce only its own specific cells. Bone, for example, could come only from the periosteum. Thus, for limb regeneration "all that is necessary in order that the process may take place is a supply of cells, capable of proliferation which contain 'bone idioplasm' and which are to multiply . . . but the *formation of a number of bones of a definite shape, arranged in a definite series,* must also be taken into account."

It is much to his credit that he recognized the problem of pattern formation – that is, of how the spatial organization of the bones was specified. This central problem was largely ignored for the next seventy years. He recognized that patterning was more than just multiplication of cells, and he rejected the idea that a mystical "invisible power – a *spiritus vector* or a *vis formativa* – may be present to direct their mode of increase and arrangement. We are nevertheless probably right in assuming . . . that the complex structures in living beings are produced merely by the agency of the forces present in the individual cells." Again, it is to his credit that he began to lay emphasis on the cellular basis of regeneration, an idea that, like pattern formation, was neglected for many years to come. His ideas about the idioplasm, however, were disproved by the findings of Chabry (1887) and Driesch (1891), who showed that half embryos could give rise to normal adults. He really had to struggle: "I nevertheless cannot help thinking that they do not in the least necessitate giving up . . . predestination," for "experiment may not always be the best guide." He chose to dismiss their findings about regulation in development as irrelevant to regeneration.

But workers on regeneration at the beginning of this century took a different view from Weismann's. The work was dominated by two Americans, T. H. Morgan and C. M. Child. Morgan was exceptionally gifted but abandoned experimental embryology for genetics quite early on; Child persisted with his work on gradients throughout his career.

Morgan's ambivalence

At the end of the nineteenth century, Sachs was using ideas similar to Bonnet's. He thought that the development of shoot buds and root buds depended on the presence of certain substances in the plant. Similar ideas were put forward by Goebel, who hinted that polarity itself is the cause that determines the direction of the flow. In discussing these ideas in relation to animal regeneration, Morgan (1903) says:

> The theory of mobile stuffs was first invented to explain the phenomenon that we call polarity. It is especially this part of the theory that I contest. . . . It is somewhat ludicrous to find that while the hypothesis of formative stuffs was first invented to explain the polarity of the pieces, the polarity is now assumed to account for the flow of stuffs."

For Goebel, following Sachs, had indeed ascribed the flow of "stuffs" to internal factors in the tissue, the direction to be determined by the polarity of the tissue itself. Morgan thought of the stuffs, in this case, being used up by the regenerating ends.

Morgan established his case with reference to experiments he made on the hydroid *Tubularia*. The stem, as well as the hydranth end, of *Tubularia* contains a red pigment, and, if the head is cut off, before a new head develops red pigment begins to appear in this region. Loeb had suggested that this pigment is a formative stuff whose accumulation at the anterior end leads to the formation of a new hydranth. Driesch (1891) had thought that the pigment might act in a quantitative manner. Morgan shows that there is no relation between the pigment and regeneration: The amount of pigment increases in small pieces, it does not decrease in the stem, and the pigment in the circulating fluid is ejected. More important than the dismissal of the red-pigment hypothesis is the rejection of the theory that the flow of formative stuffs is the basis of regeneration. For example, fastening a ligature in the middle of a long *Tubularia* stem that regenerates heads at both ends has no effect on the distal end and speeds up the regeneration of the proximal end. Therefore, the normal delay cannot be due to the distal end's using up the formative stuff and so delaying regeneration at the other end. Again, short pieces produce heads as quickly as long pieces do, showing that the amount of formative stuff is not relevant. "The stuff-hypothesis fails to explain the facts from every point of view. . . . The results certainly suggest that some physical factor enters into the problem."

The physical factor Morgan refers to is his tension hypothesis, which he put forward in 1901. Observing regeneration in animals, he is struck by the apparent resemblance of the change in form that they undergo to a process of expansion. The idea of expansion of a viscid body carries with it, of course, the idea of tension within the parts, and the return to the former condition is brought about by a release from tension and a return to a more stable condition. If by the intercalation of new material the extended condition is fixed, a new state of equilibrium will be established. (p. 271)

The best that can be said of this hypothesis is that Morgan himself recognized its weakness: "It may appear that this idea of a system of tension is too vague, that it fails to point out how the reorganization takes place, and that it gives not much more than the facts do themselves." Even so, Morgan's devotion to the hypothesis shows an extraordinary ambivalence and lack of judgment.

We have already seen Morgan's hostility to the theory of formative stuffs in his 1903 paper. Yet in his paper in 1897 on the regeneration of an earthworm he had put forward the idea of "gradients" of formative stuffs to account for the results. In this paper Morgan shows that small pieces taken from the end closer to the head, which are incapable of regenerating a head at the end that was closer to the tail, can still regenerate one quickly at the anterior end. This, he concludes, means that the inability to regenerate a head posteriorly does not depend directly on the size of the piece and that a small piece can regenerate anteriorly as rapidly as a large piece. Yet he found that the farther back the cut is made, the longer it takes for a piece to regenerate the head. Now he interprets the results in terms of formative stuffs:

The fact that a given region can begin to regenerate in one direction at once and in the other direction only after a long time, and that this power is connected with the distance of the cut surface from the anterior (or posterior) end, shows, I think, that we are dealing with something that is connected with the organization of the worm itself. Perhaps for want of a better expression we might speak of the cells of the worm as containing some sort of stuff that is more or less abundant in different parts of the body. The head stuff would gradually diminish as we pass posteriorly, and the tail stuff increase in the same direction. We should also think of this stuff in the cells as becoming active during regeneration. Where there is much of the head stuff, the cells can start sooner to regenerate anteriorly: where there is less it must increase first to a certain amount or strength before the part can begin to regenerate. I do not pretend that this explains anything at all, but the statement covers the results as they stand. (p. 582)

Lawrence (1988) refers to this statement as the first inkling of a morphogenetic gradient. It certainly is just that, and the paper even contains the concept of "thresholds" in the idea that regeneration starts only when the formative substance has increased "to a certain amount or strength."

Morgan analyzed the phenomenon of organic polarity in an important paper in 1904: "There is in every piece a gradation in the new material from before backwards that gives us the phenomenon we call polarity. With this difference, or polarity, as a basis the centripetal influence, acting from the surface inwards, determines the organization of the new part." Polarity is clearly linked to gradients, and the idea of "dominance," or organizing power, of the regenerating region is emphasized. This paper is also contemptuous of the idea of a formative force, but it is sympathetic to the idea that regeneration involves phenomena outside the physicist's experience. Nevertheless it will, he believes, be capable of a physical explanation.

A particularly important observation is almost thrown away in the paper. Morgan reports an experiment in which the tissue at the tip of a regenerating salamander limb has its polarity reversed, yet "there regenerates from the free end a new foot (and not a salamander)." He interprets this to mean that no other outcome is possible. It would be many years before this experiment was repeated and recognized as showing that regeneration in limbs always proceeds distally and that this is not an attempt to replace lost parts, since some of the structures regenerated are present (Rose, 1970).

Both in the 1901 book on regeneration, and again in 1903, Morgan is hostile to the idea of formative stuffs and favors the tension hypothesis. But in three papers in 1905 and 1906, he returns to his 1897 ideas and is explicit in relating gradients in substance to regeneration and polarity (Morgan 1905a, b, 1906). The first statement comes in a paper on the regeneration of heteromorphic tails in pieces of *Planaria* (Morgan 1905a). He is puzzled as to why the length of the piece is so important.

It is very significant I think, to find that in planarians the shortness of the piece is a factor that enters into the problem as to the character of the new part. I have suggested tentatively that this means that in *Planaria maculata* the tendency is stronger for the new structure to become a head than a tail, and that when the influence of polarity is removed a head appears on each end of short cross-pieces. In other worms, as in *Planaria simplicissima*, the tendency in certain posterior regions to produce a tail is stronger than that to produce a head, and two tails appear when the polarity is reduced or removed. Why should the length of the piece be so important a factor? Can it be that there is a greater difference, chemical or physical, between the two ends of a longer piece, so that a stronger polarity is present? In short pieces, from this point of view, the ends being near together are so much alike that the polarity is correspondingly reduced, and, under these conditions, the specification of the material of the old part is not sufficiently strong to determine the nature of the new part. These and many other equally obscure questions remain for future investigation to explain.

In the same year, studying the regeneration of *Tubularia* (Morgan 1905b), he tries to explain why an isolated piece regenerates a polyp first at the oral end and later at the opposite end. Considering both the nature of the structure developed and the timing, he writes, " We may assume that the gradation of the material is of such a kind that the hydranth forming material decreases from the apical towards the basal end." He goes on,

I have assumed that the stem of *Tubularia* is not homogeneous, but that from the hydranth to the base there is a graded difference and this gives the order or stratification of the material. The stimulus of the water acting on the free end arouses the formative changes which act in the direction of this existing material order. How does such a view differ from the old assumption of a "polarity" in the material? In my view there is no such directive force residing in the material as the term polarity suggests, but the polarity is only a name for the gradation of the material and on this as a basis the formative changes are carried out.

Here we have a clear statement of the relation between gradients and polarity.

In the 1906 paper he even writes, "The gradation is the polarity." Nothing could be more explicit. Yet just one year later, in his *Experimental Zoology* (1907), he has abandoned all such ideas. Writing of regeneration in the earthworm he asks:

What factor determines that the terminal organs are those that as first laid down in the new part. . . . What determines, then, that the new material forms in one case a head and in the other, a tail? . . . The centripetal influence acting on the new material at the anterior end determines therefore that this is a head, and acting on the new material at the posterior end determines that this is a tail. The centripetal influence is, according to my interpretation, nothing more than the tension of the outer layer of cells, and the pressure relations in general, in the rounded dome-shaped mass of new materials.

A similar line is taken concerning the behavior of hydra grafts with different polarities:

Here we meet with the conception of polarity as involved in the pressure relations. The polarity from this point of view is an expression of the graded pressure relations from one end of an organism to the other, which in turn may be an expression of the gradation of the tissues, and in turn may itself, under certain conditions, be the cause of differentiation.

His ambivalence has come full circle. Morgan has by now abandoned gradients and regeneration and is giving all his attention to genetics. But he has laid a clear basis for a gradient theory of regeneration.

Child and metabolic gradients

The gradient banner was picked up by Child, but with little acknowledgment to Morgan. From his studies on *Tubularia* Child could write, in 1907, that "provisionally, at least . . . we may regard polarity as an axial difference in the character and energy of reaction resulting from part physiological relations. . . . Polarity may be indicated by quantitative as well as by qualitative axial differences." But there is no reference to Morgan's ideas.

In the same 1907 paper Child put forward an idea that was to dominate his thinking for the next forty years – the idea of metabolic gradients. He had found that diluting the seawater they lived in caused more stolons to regenerate from the aboral (basal) end of *Tubularia* and concluded that this was due to an increase in metabolic rate and that "the original specification of the aboral end of a piece as a stolon-forming region is most firmly established and the reactions involved possess the greatest energy in the proximal regions of the stem."

A few years later, studying regeneration in *Planaria*, he concluded:

All of these facts indicate that a graded difference of some sort in the dynamic processes exists along the axis, and together with the results concerning size of pieces, the experiments show, first, that a certain minimal portion of this graded difference is necessary for the regulation of a piece into a whole, and second, that the rate of gradation along the axis differs in different regions of the body. In a later paper, however, I shall show that these factors are by no means constant, but depend, at least in large degree, upon the rate of metabolism or of certain metabolic processes.

It is this dynamic gradation along the axis, together with the complex of correlative conditions associated with it, which I regard as constituting physiological polarity. According to this idea polarity is not a condition of molecular orientation, but is essentially a dynamic gradient in one direction or different gradients in opposite directions along an axis, together with the conditions, and particularly the physiologic correlation between different parts along the axis, which must result from the existence of such a gradient or gradients. (Child, 1911)

This seems very similar to Morgan's views; to Child, however, there was a distinction:

According to the above view then, a certain minimal fraction of the axial gradient or gradients is necessary in every case for the formation of a whole. But it is necessary to call attention not merely to the existence of the axial gradient, but to the correlative factor in polarity. Morgan's hypothesis of the gradation of substances possesses certain features in common with my idea of a metabolic gradient, but the assumed gradation of substances is more stable and requires the assumption of migration; the most unsatisfactory feature of this hypothesis, however, is its failure to recognise the correlative factor in polarity. The gradation of substances alone cannot account for the fact that the same cells give rise under certain conditions to a head, under certain others to a tail, but when we consider that a gradation of any kind along the axis may give rise to a variety of correlative conditions, the phenomena become less puzzling, and we see that the directive or apparently directive feature of organic polarity is in reality a matter of physiologic correlation.

I find it difficult to understand what Child was driving at and in what substantial way his views about polarity differed from those of Morgan. He did, however, put forward one important new idea – that of dominance. His thoughts on this are summarized in his book *Individuality in Organisms*, published in 1915; one of its remarkable features is that with one exception, all the references are to his own work. He makes it clear that his ideas on biological organization come in part from an analogy he draws between the organism and the state. Order, he argues, requires authority, and in the organism this is provided by a "dominant" region that sets up a metabolic gradient.

To illustrate his conception of gradients and dominance, Child imagined a spherical mass of living protoplasm (Figure 12.1). He supposed that at one point, *a*, there is an increase in metabolic rate and that this increase spreads over the surface of the mass, decreasing in energy during transmission. The intensity is indicated in the figure by the width of the bands. Here is the first drawing of a gradient. Such gradients, Child believed, are the simplest expression of physiological unity and are the primary form of polarity and symmetry in development. In addition, the high point in the gradient, the apical region, is the dominant region in that it determines the rate of reaction at other levels.

In support of his ideas on metabolic gradients, Child used evidence from experiments from which he claimed that the susceptibility to various poisons could be used as an index of rate of metabolism of various body parts. Thus cyanides, for example, would have less effect on the regions with the highest metabolism, and these would die last. It was a crude method that tied gradients ever more tightly to metabolism: The escape was a long way off.

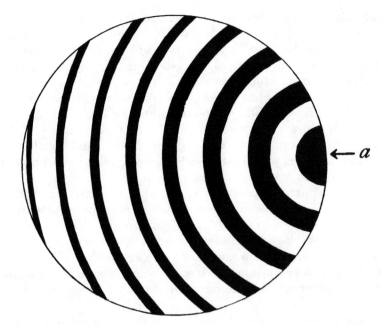

←— *a*

Figure 12.1. The first illustration of a gradient (Child, 1915). On a mass of spherical protoplasm, the point *a* is the site of an increase in metabolic rate, which spreads over the surface of the mass in a process of transmission that Child considered to be analogous to the spreading of waves in a pond. There is a decrement in intensity, indicated by the narrowing of the bands with distance from *a*. The result is a gradient in metabolic rate. (Reproduced with permission of the University of Chicago Press.)

Gradients in embryos

The concept of polarity in developing systems was first established in relation to regeneration. Although earlier workers such as Trembley, Spallanzani, and Bonnet knew that in general the end of the segment of an animal from which a head regenerates is that closest to the old head, Allman was the first to call this "polarity" (1864; reviewed in Morgan, 1901). Driesch had drawn an analogy between the polar structure of the egg and a magnet: On this basis, egg fragments should behave similarly, particularly if they are part of a "harmonious equipotential field" (1891). But Boveri (1901) pointed out that fragments of sea urchin eggs behave differently and argued that this could be explained only by a stratification of the egg. This stratification is, according to Spemann's (1938) reading of Boveri, conceived of as a gradient. The more one proceeds from the vegetative to the animal pole, the less vegetative do the strata become. At a definite level they are insufficiently vegetative to gastrulate. "Boveri . . . has even stated the idea that the most vegetative part of each egg fragment starting from a definite minimum of vegetative quality, might be a 'region of preference' from which the rest of the germ would be determined in its development." Thus Boveri (1901) is usually credited with being the first to have identified polarity with gradients. We have already seen that Morgan deserves the credit, and my reading of Boveri's paper does not easily confirm Spemann's interpretation of it. Although Boveri writes extensively about regulation and polarity, he makes no clear statement about gradients.

Child, during this period, was primarily interested in gradients in relation to regeneration, but other statements about gradients in eggs were made by Boveri (1910) and Runnstrom (1914). Boveri put forward the possibility of a gradient in relation to the development of the *Ascaris* egg. Runnstrom was more explicit, basing his ideas on his studies of bilateral symmetry in sea urchin embryos and the effect on them of different ions. He referred to neither Child nor Morgan:

During development certain power is created which transforms into "qualities." As the basis of these powers we may think of a particular specific chemical material. . . . The chemical material is directed into certain directions. These directions follow the axis of the embryo or more correctly the expression of the axes is shown in the direction of the chemical. The material is localised in a way that different parts of the embryo have different concentrations. It creates a concentration gradient. The phenomenon of polarity is an expression of the presence of this concentration gradient.

Gradients and fields

An important discovery relating to regeneration was made by Browne in 1909. She showed that the head region of hydra could, when grafted onto the body of another animal, induce the formation of a new axis. She also showed that this property was not present in a regenerating head until it had reached a certain stage in its own development. The importance of these results was not recognized in her day. It is not easy to understand how she, and those who read her paper, missed the obvious conclusion that the head acts as an inhibitor to regeneration of heads in the rest of the animal. It was only in 1926 that Rand, Bovard, and Minnich (1926) introduced the concept of inhibition. Until then, all attention had been focused on the formative dominance of the head region. They now claimed to provide proof of "dominance of the opposite sort" and referred to dominance of this kind as "inhibition." They created hydra with two heads by grafting a head from one hydra to the body column of another. After excision of the host head, only eight of eighteen hydras regenerated a new head, and this inhibition correlated with the distance between the head and the cut end: If the distance was short, inhibition occurred.

About this time (1921–3), the "field" concept was introduced into the theory of development (Huxley and de Beer, 1934). Spemann was probably the first to use it, in relation to his work on the organizer; borrowing the field concept from physics, he remarked that the organizer established a "field of organization." The idea was also introduced by Gurwitch and by Weiss, the latter applying it to the behavior of the blastema of the regenerating limb. Much later, in 1939, Weiss wrote, "The differentiation of a limb regenerate is directed and controlled by the limb field of the stump. As in ontogeny the limb field is a property of the field ... as a whole. ... A field is the condition in which a living system owes its typical organization and its specific activities." Gradients were linked to fields:

The field gradient in any given point in the field can be defined as that direction along which the field intensity falls off most rapidly. It must be borne in mind, however, that field gradients are merely convenient symbols to indicate the direction and rapidity of the decline of the resultant field action; as physical entities they are just as fictitious and non-existent as the field centre. (Weiss, 1939)

One can see that gradients and fields were still rather vague concepts.

The clearest view of gradients, some thirty years after their conception, is the brilliant synthesis by Huxley and de Beer (1934). Largely on the basis of the behavior of regenerating systems, they put forward a number of rules concerning field-gradient systems. Seven of these rules are especially relevant to this discussion: (1) The origin of polarity is to be sought in external factors, although in some cases the system is already polarized by virtue of possessing parts of the original gradient system. (2) In generation, the apical region, or head, is the first to be formed, and its formation is autonomous. (3) The newly formed apical region influences adjacent regions. (This is Child's "dominant" region.). Its influence is to establish a field. (4) The development of all other regions is dependent on influences that proceed from more apical levels. (5) At least one of the influences exerted by the more apical regions on lower levels is that of inhibition. The presence of the apical region inhibits lower levels from regenerating.

(6) If a portion of tissue comes to lie outside the field dominated by an existing apical region, a new apical region will arise. (7) The frequency or absence of regeneration and the type of structure regenerated depend on the level of the cut surface in the gradient field and the form and steepness of the gradient established. As can be seen, these rules begin to integrate the gradient concept with the properties of the apical region, which has both organizer and inhibitory activities.

In his Silliman lectures, Spemann (1938) gave a detailed critical analysis of what he called the "gradient theory." He found it difficult to understand how qualitative differences could arise from quantitative ones in the various regions: "The gradient, notwithstanding the different steepness of its single stretches must be conceived of as continuous, whereas the series of formation whose differentiation would be determined by that gradient, notwithstanding a general decrease in mass, is absolutely discontinuous."

Although there was now a useful set of rules relating to gradients and dominance in regeneration, it is striking how little attention had been given to the theoretical side of the problem, particularly the construction of working models. In this connection, Child's (1941) synthesis is disappointing. Although he recognized that the high region of a gradient is "primarily the chief dominant region," he did not see the next logical step, namely, that regulation requires the establishment of a similar high point at the cut surface. His ideas on the physiological basis of the gradient were naive. He argued, for example, that "if decrease in concentration of amount of a certain substance or substance-complex occurs in one direction along a gradient, there must be increase in concentration or amount of another substance or substances unless there is a decrease in volume." He, like his contemporaries, was obsessed with metabolism, and it was only when this paradigm was replaced with that of information transfer that new ideas emerged. Although he recognized that the gradient system can be viewed as a physiological coordinate system, he did not pursue this thought.

In contrast to the rather vague ideas of Child, Spiegelman (1945) and Rose (1952) put forward specific models. Spiegelman pointed out that one of the features of regeneration is that the potentiality for forming a particular part of the pattern is present in a larger area than actually forms the part. He argued for a "principle of limited realization," which he suggested must involve two distinct mechanisms. The first is the suppression of the realization of potentialities, and the second provides for differences between parts of the system. If, for example, two parts of the system are capable of forming a particular region, and only one does so, then it is essential to have both a difference between these two parts and the suppression of the development of one part by the other. He stressed that a difference, as manifested by a gradient in some property, would not in itself provide an adequate mechanism.

Rose's model (1952) was even more explicit; in fact, it was the most explicit model of its time. He suggested that the genesis of a pattern could be a consequence of a hierarchy of self-limiting reactions, together with the spread of inhibition from one differentiating region to others farther down the gradient. The gradient provides the basis of the rate of differentiation, the reactions proceeding fastest at the high point of the gradient. The reaction that predominates is the one closest to the top of the hierarchy, which has not yet been inhibited. Each reaction is self-limiting by virtue of the inhibitor it produces. The model actually generated a pattern and, though not based directly on the gradient theory, required at least a "high point," and this implied graded differences. (For some criticisms of the model, see Wolpert, 1968.) In a later version it was assumed that the transport of the inhibitory substances is polarized in a head-to-tail direction (reviewed in Webster, 1971).

The modern period

It is too soon to write a history of the modern period, which we can take as beginning in the 1960s. Webster and I would like to believe that our studies of hydra, which were based on and extended those of Browne (1909) and Rand, Bovard, and Minnich (1926), made a useful contribution to gradient theory (Webster, 1971). We showed (Webster and Wolpert, 1966) that the time required for determination of the head end, as assayed by Browne's technique, was very rapid compared to the time required for morphological differentiation of the head, and there was an axial gradient in the time required for head regeneration. More important, our experiments distinguished between two gradients: a gradient in head inhibitor and a gradient in the threshold to inhibition (Webster, 1971). Our key experiment demonstrated that regions from just below the head when grafted near the foot end of the animal could form heads, but when grafted near the head they were inhibited. These experiments were elaborated by McWilliams (reviewed in Bode and Bode, 1984).

On the theoretical side, the problem of regeneration in hydra was considered in relation to the French-flag problem (Wolpert, 1968). That problem is: Given a line of cells that can be blue, red, or white, how should they communicate with each other so as to form a French flag that is one-third red, one-third white, and one-third blue, and continue to do so even when parts are removed? A formal solution was proposed in terms of positional information, which freed gradients from metabolism and gave them a more rigorous definition and conceptual framework (Wolpert, 1969, 1971). Cells in developing and regenerating systems can be thought of as having their position specified, as in a coordinate system. Their pattern of differentiation then results from the cells' interpreting their positional value according to their genetic constitution and developmental history. A "field" can thus be thought of as a group of cells sharing a common coordinate system. (For a general history of the idea of gradients in development, see Wolpert, 1981).

From the point of view of positional information, regeneration is the reestablishment of positional values. Morgan (1901) distinguished between *morphallaxis* and *epimorphosis* in regeneration. Morphallaxis (in hydra regeneration, for example) involves not growth but the respecification of existing tissue. In terms of positional information, this means the establishment of "boundary regions" or "reference regions" as a first step (Wolpert, 1971). Thus the head end of the hydra, Child's "dominant region," is the boundary with respect to which the positions of the other cells are specified. In contrast to epimorphic regeneration, there is growth, and new positional values are generated from the growing tissue. There is no need to specify new boundary regions. By far the most imaginative application of these ideas is the polar coordinate model proposed by French, Bryant, and Bryant (1976). They suggested that the positional values of the cells are specified as in a polar coordinate system and that when grafts are made, thereby disrupting the system, new positional values are intercalated to smooth out differences. In addition, the regeneration of structures in a distal direction can be related to this intercalation. An early key experiment investigating these processes was by Bohn (1970). He showed that the tibia of the cockroach behaves as if there is a graded set of properties and that, after partial limb resection intercalation occurs to restore this gradient to its original slope. (For recent ideas on gradients, see Wolpert, 1989).

A theoretical approach that has attracted much attention uses the reaction-diffusion model proposed by Turing. In a brilliant paper (1952) he showed that

a system of chemical substances, called morphogens, reacting together and diffusing through a tissue, is adequate to account for the main phenomena of morphogenesis. Such a system, although it may originally be quite homogeneous, may later develop a pattern or structure due to an instability of the homogeneous equilibrium, which is triggered off by random disturbances.

Although Turing was primarily interested in generating spatial patterns and did not consider regeneration at all, such reaction-diffusion systems have been explored as a possible basis for regeneration, since they provide a self-organizing chemical system for generating gradients that are capable of regeneration (Gierer, 1981).

Conclusions

The idea that gradients play a key role in regeneration was clearly stated by Morgan at the beginning of the century, and Child devoted most of his life to studying gradients; yet by the 1950s the understanding of gradients had made little progress. Few clearly worked out models or mechanisms existed, and there was a notable lack of quantitative analysis. It seems likely that the failure to adopt a quantitative and mechanistic approach accounts for the lack of progress. Little attempt was made to model how gradients are set up or regulated. Any of the grafting experiments carried out on hydra in the 1960s and 1970s could have been done at the beginning of the century. So the question is, Why were they not done then? One reason was the absence, in studies of regeneration, of a commitment to quantitative studies and to the construction of models. But probably more important was the fact that gradients were considered solely within the conceptual framework of metabolism (Wolpert, 1981). Metabolism was the paradigm within which biological mechanisms were considered, and the shift from metabolism to the information paradigm occurred only with the coming of molecular genetics. Brenner has pointed out that before Sanger showed that proteins are a specific linear sequence of amino acids, the questions were all about where the energy for making proteins comes from, and it was a major change to start asking, instead, what information determines the sequence of amino acids (Wolpert and Richards, 1988). The shift to an informational framework undoubtedly influenced thinking about regeneration. Instead of concentrating on metabolism, thinking became focused on the information necessary for specifying the lost parts, such as the head end in hydra. Regeneration became, as it should be, concerned with cell–cell interactions.

Finally, there remains the puzzle of Morgan's ambivalence. He undoubtedly was the first to put forward the concept of gradients in relation to regeneration, particularly to polarity; yet he kept switching to a rather vague mechanical model. It is interesting to speculate what progress there would have been if he had confined his attention to regeneration. He must have realized that the hope of real progress lay elsewhere. It is a measure of the correctness of his judgment that he turned to genetics and won a well-deserved Nobel prize.

REFERENCES

Bode, P. M., and H. R. Bode. 1984. Patterning in hydra. In *Pattern Formation*, ed. G. Malacinski, pp. 213–43. New York: Macmillan.

Bohn, H. 1970. Interkalare Regeneration und segmentale Gradienten bei den Extremitaten von *Leucophaea*-Larven (Blattaria). I. Femur und Tibia. *Wilhelm Roux Arch. EntwMech. Org.*, 165: 303–41.

Boveri, T. 1901. Über die Polarität des Seeigeleies. *Verk. Phys-med. Ges. Wurzburg*, N.F. 31, Vol. 34, pp. 145–70.

1910 Die Potenzen der *Ascaris*-Blastomeren bei abgeänderter Furchung zugleich ein Beitrag zur Frage qualitätive rungleicher Chromosomenteilung. In *Festschrift zum 60. Geburtstag von R. Hertwig*, 3: 131–220. Jena: Fischer.

Browne, E. H. 1909. The production of new hydranths by the insertion of small grafts. *J. Exp. Zool.*, 7: 1–23.

Chabry, L. M. 1887. Contribution à l'embryologie normale tératologique des ascides simples. *Journal d'anatomie et de la physiologie normales et pathologiques de l'homme et des animaux*, 23: 167–321.

Child, C. M. 1907. An analysis of form regulation in *Tubularia* I. *Arch. Entw. Mechan.*, 23: 396–414.

1911. Studies on the dynamics of morphogenesis and inheritance in experimental reproduction. I. The axial gradient in *Planaria dorotocephala* as a limiting factor in regeneration. *J. Exp. Zool.*, 10: 265–320.

1915. *Individuality in Organisms*. Chicago: University of Chicago Press.

1941. *Patterns and Problems of Development*. Chicago: University of Chicago Press.

Driesch, H. 1891. *Entwicklungmechanische Studien. Zeit. f. Wiss. Zoologie*, 53.

French, V., P. J. Bryant, and S. V. Bryant. 1976. Pattern regulation in epimorphic fields. *Science*, 193: 969–81.

Gierer, A. 1981. Some physical, mathematical and evolutionary aspects of biological pattern formation. *Philos. Trans. R. Soc. B*, 295: 425–617.

Huxley, J. S., and G. R. de Beer. 1934. *The Elements of Experimental Embryology*. Cambridge: Cambridge University Press.

Lawrence, P. A. 1988. Background to bicoid. *Cell*, 54: 1–2.

Morgan, T. H. 1897. Regeneration in *Albobophora foetida*. *Roux Arch. EntwMech.*, 5: 570–86.

1901. *Regeneration*. New York: Macmillan.

1903. The hypothesis of formative stuffs. *Torrey Botanical Club Bulletin*, 30: 206–13.

1904. An analysis of the phenomena of organic polarities. *Science*, 20: 742–8.

1905a. Regeneration of heteromorphic tails in posterior pieces of *Planaria simplicissima. J. Exp. Zool.*, 1: 385–93.

1905b. An attempt to analyse the phenomena of polarity in *Tubularia. J. Exp. Zool.*, 1: 589–91.

1906. "Polarity" considered as a phenomenon of gradation of materials. *J. Exp. Zool.*, 2: 495.

1907. *Experimental Zoology.* New York: Macmillan.

Rand, H. W.; J. F. Bovard; and D. E. Minnich. 1926. Localization of formative agencies in hydra. *Proc. Natl. Acad. Sci.*, 12: 565–70.

Rose, S. M. 1952. A hierarchy of self-limiting reactions as the basis of cellular differentiation and growth control. *Am. Nat.*, 86: 337.

1970. *Regeneration.* New York: Appleton-Century-Crofts.

Runnstrom, J. 1914. Analytische Studien über Seeigelentwicklung I. *Arch. Entwick.*, 40: 526–64.

Spemann, H. 1938. *Embryonic Development and Induction.* New Haven: Yale University Press.

Spiegelman, S. 1945. Physiological competition as a regulatory mechanism in morphogenesis. *Q. Rev. Biol.*, 20: 121–46.

Turing, A. M 1952. The chemical basis of morphogenesis. *Philos. Trans. R. Soc. B*, 237: 37–84.

Webster, G. 1971. Morphogenesis and pattern formation in hydroids. *Biological Reviews*, 46: 1–46.

Webster, G., and L. Wolpert. 1966. Studies on pattern regulation in hydra. I. Regional differences in time required for hypostomal determination. *J. Embryol. Exp. Morphol.*, 16: 91–104.

Weismann, A. 1893. *The Germ-Plasm: A Theory of Heredity.* London, Walter Scott.

Weiss, P. 1939. *Principles of Development.* New York: Holt, Rinehart & Winston.

Wolpert, L. 1968. The French flag problem: A contribution to the discussion on pattern development and regulation. In *Towards a Theoretical Biology*, ed. C. H. Waddington, pp. 125–33. Edinburgh: Edinburgh University Press.

1969. Positional information and the spatial pattern of cellular differentiation. *J. Theoret. Biol.*, 25: 1–47.

1971. Positional information and pattern formation. *Curr. Top. Devel. Biol.*, 6: 183–211.

1981. Gradient, position and pattern: A history. In *A History of Embryology*, ed. T. J. Horder, J. A. Witkowski, and C. C. Wylie, pp. 347–61. Cambridge: Cambridge University Press.

1989. Positional information revisited. *Development*, 107 (suppl.): 3–12.

Wolpert, L., and A. Richards. 1988. *A Passion for Science.* Oxford: Oxford University Press.

Index

219